Nelson Mathematics 7

Workbook

Series Authors and Senior Consultants
Marian Small • Mary Lou Kestell

Senior Author
David Zimmer

Workbook Author
Christy Hayhoe

NELSON

NELSON

Nelson Mathematics 7
Workbook

Series Authors and Senior Consultants
Marian Small, Mary Lou Kestell

Senior Author
David Zimmer

Workbook Author
Christy Hayhoe

Director of Publishing
Beverley Buxton

Acquisitions Editor
Colin Garnham

Project Manager, K–8
David Spiegel

Senior Program Manager
Shirley Barrett

Developmental Editor
Christy Hayhoe

Executive Managing Editor, Development & Testing
Cheryl Turner

Executive Managing Editor, Production
Nicola Balfour

Workbook Author
Christy Hayhoe

Senior Production Editor
Susan Skivington

Copyeditor
Susan McNish

Editorial Assistant
Amanda Davis

Senior Production Coordinator
Sharon Latta Paterson

Production Coordinator
Franca Mandarino

Creative Director
Angela Cluer

Art Director
Ken Phipps

Art Management
Suzanne Peden

Illustrators
Deborah Crowle, Steven Corrigan

Interior Design
Suzanne Peden

Cover Design
Ken Phipps, Peggy Rhodes

Cover Image
Corbis/Magna

Composition
Janet Zanette

Library and Archives Canada Cataloguing in Publication

Nelson mathematics 7.
Workbook / Zimmer ... [et al.].

ISBN 0-17-626994-0

1. Mathematics—Problems, exercises, etc..

I. Zimmer, David II. Title. III. Nelson mathematics seven.

QA107.2.N44 2005 Suppl. 3
510   C2005-903592-7

Contents

Message to Parent/Guardian

This workbook contains questions for each lesson in your child's textbook *Nelson Mathematics 7*. The questions in the workbook are similar to the ones in the text, so they should look familiar to your child. The lesson Goal and the At-Home Help on each page will help you to provide support if your child needs it.

At the end of each chapter are two pages of multiple-choice questions called "Test Yourself." This is an opportunity for you and your child to see how well she or he understands.

You can help your child to explore and understand math ideas by making available some commonly found materials, such as

- string, scissors, and a ruler (for measurement)
- counters such as bread tags, toothpicks, buttons, or coins (for number operations and patterns)
- packages, cans, toothpicks, and modelling clay (for geometry)
- grid paper, magazines, and newspapers (for data management)
- board game spinners, dice, and card games (for probability)

You might also encourage your child to use technology if it is available, such as

- a calculator (for exploring number patterns and operations)
- a computer (for investigating the wealth of information that exists on the Internet to help people learn and enjoy math)

Visit the Nelson Web site at **www.mathk8.nelson.com** to find out more about the mathematics your child is learning.

It's amazing what you can learn when you look at math through your child's eyes! Here are some things you might watch for.

Checklist
- ☑ Can your child clearly explain her or his thinking?
- ☑ Does your child check to see whether an answer makes sense?
- ☑ Does your child persevere until the work is complete?
- ☑ Does your child connect new concepts to what has already been learned?
- ☑ Is your child proud of what's been accomplished so far?

1.1 Using Multiples

▶ **GOAL** Identify multiples, common multiples, and least common multiples of whole numbers.

1. The multiples of 2 and 5 are given below.

   2, 4, 6, 8, 10, 12, 14, 16, 18, 20, …

   5, 10, 15, 20, 25, 30, 35, 40, …

   a) List two common multiples of 2 and 5.
   ten and twenty

   b) What is the LCM of 2 and 5?
   _10_

2. List the first 10 multiples for each number.

   a) 3: _3,6,9,12,15,18,21,24,27,30_

   b) 4: _4,8,12,16,24,28,32,36,40_

   c) 7: _7,14,21,28,35,42,49_

At-Home Help

A **multiple** is the product of a whole number (for example, 1, 2, 3, …) when multiplied by any other whole number. The **least common multiple** (LCM) is the lowest multiple that two or more numbers have in common.

For example:
- The multiples of 2 are 2, 4, 6, 8, 10, 12, 14, 16, 18, …
- The multiples of 3 are 3, 6, 9, 12, 15, 18, 21, …
- 6, 12, and 18 are multiples of both 3 and 2. The LCM of 3 and 2 is 6.

3. Continue the patterns you started in question 3 to find the LCM of 3, 4, and 7.

4. Find the LCM of each set of numbers.

   a) 2 and 4   d) 2, 4, and 12

   b) 5 and 6   e) 3, 6, and 8

   c) 8 and 10   f) 3, 10, and 15

5. Yuki gets her hair cut every three months. She buys a new shirt every four months. How many times a year does Yuki have a new haircut and a new shirt in the same month?

   hair cut 3,6,9,12,
   shirt 4,8,12,
   Since Yuki get a new sirt she will bey both once a year / every 12 months

6. Indira, Bonnie, and Chang all work at the same pet store. Indira works every other day. Bonnie works every fourth day. Chang works every fifth day. Today they all worked together. How many days will it be until they work together again?

   Indira 2,4,6,8,10,12,14,16,18,20   They will alwork together on
   Bonnie 4,8,12,16,20   in 20 days
   Chang 5,10,15,20

1.2 A Factoring Experiment

▶ **GOAL: Identify factors of numbers.**

1. In the chart to the right, numbers up to 40 that have a factor of 3 are highlighted.

1	2	3	4	5	6	7	8	9	10
11	12	13	14	15	16	17	18	19	20
21	22	23	24	25	26	27	28	29	30
31	32	33	34	35	36	37	38	39	40
41	42	43	44	45	46	47	48	49	50
51	52	53	54	55	56	57	58	59	60
61	62	63	64	65	66	67	68	69	70
71	72	73	74	75	76	77	78	79	80
81	82	83	84	85	86	87	88	89	90
91	92	93	94	95	96	97	98	99	100

   a) What pattern do you see?

   _____

   _____

   b) Use the pattern to help you complete the chart up to 100.

   c) The number 3 is one factor of 27. What is another factor of 27? _____

2. In the chart to the right, numbers up to 40 with a factor of 4 are highlighted.

   a) What pattern do you see? _____

   b) Use the pattern to help you complete the chart up to 100.

   c) The number 4 is one factor of 32. List five more factors of 32.

   _____

1	2	3	4	5	6	7	8	9	10
11	12	13	14	15	16	17	18	19	20
21	22	23	24	25	26	27	28	29	30
31	32	33	34	35	36	37	38	39	40
41	42	43	44	45	46	47	48	49	50
51	52	53	54	55	56	57	58	59	60
61	62	63	64	65	66	67	68	69	70
71	72	73	74	75	76	77	78	79	80
81	82	83	84	85	86	87	88	89	90
91	92	93	94	95	96	97	98	99	100

3. The number 24 is highlighted in both charts above. This means that 3 and 4 are both factors of 24.

   a) Use the charts to find five more numbers with 3 and 4 as factors.

   _____

   b) Complete the equations.

   $1 \times$ ___ $= 24$ $3 \times$ ___ $= 24$ $6 \times$ ___ $= 24$ $12 \times$ ___ $= 24$

   $2 \times$ ___ $= 24$ $4 \times$ ___ $= 24$ $8 \times$ ___ $= 24$ $24 \times$ ___ $= 24$

   c) The number 24 has eight factors in total. What are they?

   _____

1.3 Factoring

▶ **GOAL:** Determine factors, common factors, and greatest common factors of whole numbers.

1. **a)** You can find factors by dividing. Fill in the blanks to find the factors of 12.

   $12 \div 1 = \underline{12}$ $12 \div 3 = \underline{4}$ $12 \div 6 = \underline{2}$

   $12 \div 2 = \underline{6}$ $12 \div 4 = \underline{3}$ $12 \div 12 = \underline{0}$

   b) How do you know that 5, 7, 8, 9, 10, and 11 are not factors of 12?

   I know the are not factors because they do not mult-ail also if I divide them it will give me a 8.0 number

2. **a)** Fill in the missing factors in this factor rainbow.

   1 ? 5   ? 15 45

   b) What are the factors of 45? ___ 4.c) ___

   c) Make your own factor rainbow for the number 50.

   1 2 5 10 25 50

3. Find the factors for each number.

   a) 15   1×15   3×5

   b) 40   1×40   2×20   4×10   5×8

   c) 54   1×54   3×15   5×145

   d) 72   1×72   2×36   3×24

4. Use your answers from question 3 to find the GCF for each pair.

   a) 15 and 40: ___ 5 ___   **b)** 40 and 72: ___ 8 ___

5. Find the GCF of each pair of numbers.

   a) 28, 84   28

   b) 65, 39   13

   c) 75, 45   15

6. How many ways can you stack 77 books into equal piles?   4 times ✓

   77   1×77   7×11

1.4 Exploring Divisibility

▶ **GOAL: Use divisibility rules to identify factors of numbers.**

1. Which of these numbers are divisible by 2? Use the divisibility rules to help you.

   a) 243: _____

   b) 634: _____

   c) 937: _____

   d) 58 930: _____

At-Home Help

These rules can help you find factors without having to divide.
- Even numbers are divisible by 2.
- If the last digit is 0 or 5, the number is divisible by 5.
- If the last digit is 0, the number is divisible by 10.
- A number is divisible by 3 if the sum of its digits is a multiple of 3.
- A number is divisible by 9 if the sum of its digits is a multiple of 9.

2. Which of these numbers are divisible by both 5 and 10? Use the divisibility rules to help you.

   a) 943 720: _____

   b) 345: _____

   c) 57 382: _____

   d) 8880: _____

3. Which of these numbers are divisible by 3? Use the divisibility rules to help you.

   a) 7364: _____

   b) 1335: _____

   c) 2352: _____

   d) 6342: _____

4. Which of these numbers are divisible by 9? Use the divisibility rules to help you.

   a) 6966: _____   b) 8366: _____   c) 39 015: _____

5. In the following chart, shade the statements that are false to find the hidden message.

4185 is divisible by 9	7655 is divisible by 9	9965 is divisible by 5	3345 is divisible by 10	51 315 is divisible by 3 and 5	432 is divisible by 5
832 is divisible by 2	1863 is divisible by 3	88 is divisible by 2	777 is divisible by 5	7332 is divisible by 3	3740 is divisible by 5 but not 10
3548 is divisible by 2	6740 is divisible by 3 and 10	810 is divisible by 2, 5, and 9	8324 is divisible by 10	270 is divisible by 3 and 2	4766 is divisible by 5
38 920 is divisible by 10 but not 3	553 is divisible by 2	5416 is divisible by 2	6615 is divisible by 5 but not 3	1485 is divisible by 3 and 9	334 is divisible by 9

1.5 Powers

▶ **GOAL: Use powers to represent repeated multiplication.**

Handwritten in margin:
$\begin{array}{r} 1 \\ 16 \\ \times\ 2 \\ \hline 32 \\ \times\ 2 \\ \hline 64 \end{array}$

1. Use powers to represent each multiplication. Then calculate each product.

   a) $3 \times 3 \times 3 =$ _$3^3 \div 3 \times 3 \times 3 = 27$_

   b) $6 \times 6 =$ _$6^2 = 6 \times 6 = 36$_

   c) $2 \times 2 \times 2 \times 2 \times 2 \times 2 =$ _$2^6 = 2 \times 2 \times 2 \times 2 \times 2 \times 2 = 64$_

   handwritten: 4 8 16 32 64

   d) $10 \times 10 \times 10 \times 10 \times 10 =$ _$10^5 = 10 \times 10 \times 10 \times 10 \times 10$_

> **At-Home Help**
>
> A **power** is a numerical expression that shows repeated multiplication.
>
> For example, in the expression 5^4, the number 5 is being multiplied by itself four times.
> $5^4 = 5 \times 5 \times 5 \times 5$
>
> Remember that 5^4 is not the same as 5×4.

2. Write each power as a repeated multiplication. Then calculate.

   a) $10^4 =$ _____

   b) $5^5 =$ _____

   c) $2^8 =$ _____

   d) $1^7 =$ _____

3. Which equations are not true? Write correct versions where needed.

   Handwritten in margin:
   $\begin{array}{r} 10 \\ \times 10 \\ \hline 1000 \end{array}$

   a) $2^4 = 2 \times 2 \times 2 \times 2$ _____

   b) $3^3 = 3 \times 3$ _____

   c) $4^2 = 16$ _____

   d) $6^3 = 18$ _____

4. Write each number as a power.

   a) $25 =$ _5^2_   b) $64 =$ _____   c) $8 =$ _____   d) $27 =$ _____

5. Kwami had two large sheets of paper. He cut each piece of paper in half, then he cut the smaller pieces of paper in half two more times.

   a) Write this as a power. _____

   b) How many pieces of paper does Kwami have now? _____

6. Fill in the chart. The first row is done for you.

Power	Base	Exponent	Meaning	Product
3^2	3	2	3×3	9
2^3				
	6	3		.
			4×4	
	3			81

Copyright © 2006 by Nelson Education Ltd.

Chapter 1: Factors and Exponents **5**

1.6 Square Roots

▶ **GOAL: Determine the square roots of perfect squares.**

1. Write down the area of each square, and below it write the square root of that number. The first one is done for you.

   a) The area is 4 square units.

   The length of each side is 2 units.

   $\sqrt{4} = 2$

   b) The area is ___9___ square units.

   The length of each side is ___3___ units.

   $\sqrt{9} = $ ___3___

   c) The area is ___4 16___ square units.

   The length of each side is ___4___ units.

   $\sqrt{16} = $ ___4___

   d) The area is ___10___ square units.

   The length of each side is ___5___ units.

   $\sqrt{25} = $ ___5___

At-Home Help

To help you find the square root of a number, try comparing it to square roots of numbers you know. You can also use the last digit of the number to help you discover what the last digit of the answer must be. For example:

Question: Find the square root of 361.

Solution:

$10^2 = 10 \times 10 = 100$ (Too low)

$20^2 = 20 \times 20 = 400$ (Too high)

$1 \times 1 = 1$ or

$9 \times 9 = 81$

The answer is between 10 and 20. It is closer to 20 than 10. The last digit of the number will be 1 or 9. The answer is 11 or 19. Multiply to check:

$11 \times 11 = 121$ (Too low)

$19 \times 19 = 361$

2. Use mental math to find the square root of each number.

   a) $\sqrt{36} = $ ___6___   b) $\sqrt{64} = $ ___8___   c) $\sqrt{49} = $ ___7___

3. A restaurant has a square floor with an area of 324 m². Sketch the restaurant floor. What are its dimensions?

4. Guess and test to find the square root of each number.

   a) 441   c) 625   e) 6724

   b) 484   d) 1024   f) 1681

1.7 Order of Operations

▶ **GOAL: Apply the rules for order of operations.**

1. Calculate.

   a) $6 + 2 \times 5$

   b) $(6 + 2) \times 5$

   c) $14 - 6 \div 2$

   d) $(14 - 6) \div 2$

   e) $8 + 12 \div 4 - 2$

   f) $(8 + 12) \div 4 - 2$

   g) $(8 + 12) \div (4 - 2)$

   h) $8 + 12 \div (4 - 2)$

   i) $4 + 3^2 \times 2 - 1$

   j) $(4 + 3)^2 \times 2 - 1$

   k) $4 + (3 \times 2)^2 - 1$

   l) $4 + 3^2 \times (2 - 1)$

2. Place brackets in the following to make each statement true.

   a) $3 + 2 \times 4 = 20$

   b) $20 - 6 \times 2 + 1 = 2$

   c) $4 \times 2 + 3 + 1 = 21$

   d) $4 \times 2 + 3 + 1 = 24$

   e) $4 + 2 \times 3 - 2 = 6$

   f) $7 - 2^2 + 3 = 28$

   g) $2 + 2 + 2^2 = 36$

   h) $8 - 3 \times 2 - 1 = 5$

3. Calculate.

   a) $6 + 5 \times (4 - 1)$

   b) $18 \div (12 - 3) + 5$

   c) $(12 + 20) \div (9 \times 2 - 2)$

   f) $9 \times 4 + 4 \div 2$

   g) $(4 + 1 + 2)^2 - (3 \times 4 + 2)$

   h) $(1 \times 1 + 1 \times 1)^3 \times (2 - 1 \times 1)^2$

d) $3^2 + 4 \times 2$

i) $3^2 - 8 \div 4 + (6 - 2)^2$

e) $7 - 2 \times 3 + 4$

j) $(54 - 7^2)^2$

4. Romona, Miguel, and Fawn each got a different answer when they solved the following question:

   $2 \times 4^2 - 2 \div 2$

   Romona: 15   Miguel: 63   Fawn: 31

   Who is correct? Show your work.

5. Circle the expression with the greatest value.

   a) $6 \times 3 \div 2$ **b)** $6 \div 3 \times 2$ **c)** $16 \div 4 \div 2$ **d)** $2 \times 6 \times 4$

6. Circle the expression with the lowest value.

   a) $4 \times 3 + 2 - 12 \div 2$ **b)** $(4 \times 3 + 2 - 12) \div 2$ **c)** $4 \times (3 + 2) - 12 \div 2$

7. Fill in the blanks with >, <, or = to make the statements true.

   a) $3 \times 2 - 5$ _____ $5 \times 2 - 3$ **e)** $36 \div 12 \div 3$ _____ $24 \div 12 \div 2$

   b) $3^2 - 2$ _____ $(3 - 2)^2$ **f)** $23 - 12 \div 6$ _____ $3 + 6 \times 3$

   c) $(8 + 7) \times 5$ _____ $8 + 7 \times 5$ **g)** $5 \times 9 \div 3$ _____ $8 + 4 \times 8$

   d) $7 + 2 \times 4$ _____ $5 \times 2 + 6$ **h)** $5 \times 5 \div 5$ _____ $5 \div 5 \times 5$

8. Ravi's shirts were in three piles of four shirts each, and his pants were in two piles of six pairs each. How many articles of clothing does Ravi have in total?

   a) Write the problem as a mathematical expression.

   b) Calculate. How many articles of clothing does Ravi have? _____

1.8 Solve Problems by Using Power Patterns

▶ **GOAL:** Use patterns to solve problems with powers.

1. a) Look at the equations below. What pattern do you see?

_____

$10^1 = 10$

$10^2 = 100$

$10^3 = 1000$

$10^4 = 10\ 000$

b) Use the pattern you saw in part (a) to predict the number of zeros in 10^{100}.

_____

2. Predict the last digit for 5^{16}. Find 5^1, 5^2, 5^3, 5^4, ..., and look at the last digit each time.

At-Home Help

Looking at patterns can help you make predictions. For example, watch for repeated last digits, such as **2, 4, 8, 16, 32, 64, 128, 256**, and so on. Try making a chart to record the pattern you see, like this:

Column 1	Column 2	Column 3	Column 4
$2^1 = 2$	$2^2 = 4$	$2^3 = 8$	$2^4 = 16$
$2^5 = 32$	$2^6 = 64$	$2^7 = 128$	$2^8 = 256$

You can extend this chart and use the pattern to make predictions about the last digit of larger numbers.

3. Notice the pattern below.

$1 = 1$

$1 + 2 + 1 = 4$

$1 + 2 + 3 + 2 + 1 = 9$

a) Find the sum using the pattern.

$1 + 2 + 3 + 4 + 3 + 2 + 1 = $ _____

b) Determine a rule for finding the sum.

$1 + 2 + 3 + 4 + 5 + 4 + 3 + 2 + 1 = $ _____

_____

c) Use your rule to find the sum.

$1 + 2 + 3 + 4 + 5 + 6 + 7 + 8 + 9 + 8 + 7 + 6 + 5 + 4 + 3 + 2 + 1 = $ _____

Test Yourself

1. List the first six multiples of each number.

   a) 2: 2,4,6,8,10,12

   b) 8: 8,16,24,32,40,48

   c) 10: 10,20,30,40,50,60

   d) 12: 12,24,36,48,60,72

2. Is the first number a multiple of the second?

   a) 56, 8: _____   c) 168, 4: _____

   b) 232, 3: _____   d) 3215, 5: _____

3. Find the LCM of each pair of numbers.

   a) 5, 6   c) 6, 8

   b) 4, 7   d) 4, 10

4. Mrs. Simpson waters her geraniums once every three days, and she waters her other house plants once a week. How often does Mrs. Simpson water all the plants on the same day?

   Geraniums- 3,6,9,12,15,18,21

   other - 7,14,21   Mrs. Simpson will water both plants on the 21 day.

5. Find all the factors of each number.

   a) 36: 1,2,3,6,12,18,36

   b) 42: 1,2

   c) 68: _____

   d) 81: _____

   3,6,9,12,15,18,21,24,27,30,33,36

6. Find the GCF of each pair of numbers.

   a) 16, 24   c) 48, 120

   b) 39, 14   d) 28, 70

7. There are 84 people going camping.

   a) A bus can take 42 people. How many buses would be needed? _____

   b) A van can take 12 people. How many vans would be needed? _____

   c) A car can take 4 people. How many cars would be needed? _____

8. Ninety-six (96) basketball players are going to a tournament. The team wants to use as few vehicles as possible. What combination of vehicles should they use? (Use the information from question 7 to answer this question.)

9. Mr. Singh's and Mrs. Jackson's gardens are separated by a fence. The area of Mr. Singh's garden is 36 m², and the area of Mrs. Jackson's garden is 20 m².

   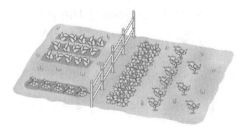

   a) What are the possible dimensions of each garden?

   b) What is the greatest length the fence could be?

2⟌36

10. Is 89 718 divisible by the number?

a) 2: _____Y_____   d) 5: _____N_____

b) 3: _____N_____   e) 6: _____N_____

c) 4: _____Y_____   f) 7: _____N_____

11. Calculate.

a) 2^5

b) 5^3

c) 10^4

12. Express as a power.

a) 27: _____   b) 128: _____

c) 1 000 000: _____

13. A square field has an area of 289 m². What are its dimensions?

14. Use mental math to find the square root of each number.

a) 100   c) 40 000

b) 10 000   d) 9 000 000

15. Calculate each expression.

a) $8 \times 4 - 2$

b) $4 + 2 \times 9 - 3^2$

c) $(3 + 2)^2 - (5 - 3)^2$

d) $5 + 10^2 + 3 \times 10^3 + 7 \times 200$

16. Determine the last digit of each power using patterns.

a) 3^{16} c) 5^{19}

b) 2^{35} d) 4^{11}

17. You can find the volume of a box by multiplying its height by its width by its depth.

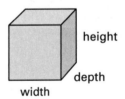
height
depth
width

You are given a box in the shape of a cube. Each edge measures 5 cm.

a) Write an expression for the volume of the box using a power.

_____

b) Calculate the volume of the box.

Exploring Ratio Relationships

▶ **GOAL:** Explore equivalent ratios.

1. Which of these pairs are similar?

   a) 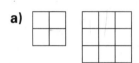   b)

At-Home Help

A **ratio** is a way to compare two or more numbers. For example, you can write the comparison of 5 girls to 4 boys in three ways:

5:4   5 to 4 $\frac{5}{4}$

Similar rectangles have the same shape, but not necessarily the same size. The ratio of width to length of one rectangle will be equivalent to the width-to-length ratio for a similar rectangle. For example:

width:length = 1:3

width:length = 2:6
$= \frac{2}{2} : \frac{6}{2}$
= 1:3

2. a) Use the grid on the right to draw two rectangles that are similar to the one on the left.

   2:4

   b) Write width-to-length ratios for the two rectangles you drew. How do they compare to the ratio of the rectangle above?

   _____

   _____

3. There are five students for every tutor in after-school classes at Middletown School.

   a) Use the grid to draw a rectangle with the number of students as the length, and the number of tutors as the width.

   b) Use the same grid to draw a rectangle with the number of students in two classes as the length, and the number of tutors in two classes as the width.

   c) Are the two rectangles you drew similar? _____

   d) What is the student-teacher ratio for one class? _____

   e) What is the student-teacher ratio for two classes? _____

   f) Are these two ratios equivalent? Explain why or why not.

   _____

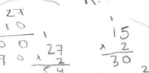

2.2 Solving Ratio Problems

▶ **GOAL:** Compare quantities using ratios, and determine equivalent ratios.

1. Write three equivalent ratios for each.

   a) 2:1 = _____

   b) 32:4 = _____

   c) 27:15 = $\frac{27}{15} = \frac{54}{30}$ | $\frac{27}{15} = \frac{135}{75}$ | $\frac{27}{15} = \frac{270}{150}$

   d) $\frac{10}{3}$ = $\frac{10}{3} = \frac{100}{30}$ | $\frac{10}{3} = \frac{50}{15}$ | $\frac{10}{3} = \frac{20}{6}$ keys,10,5,2

2. Find the missing term in each proportion.

   a) $\frac{4}{8} = \frac{1}{}$ key:

   c) $\frac{2}{7} = \frac{6}{21}$ key = 3

   b) $\frac{3}{5} = \frac{6}{10}$ key=2

   d) 42:18 = 7:3 key = 6

At-Home Help

• **Equivalent ratios** are two or more ratios that represent the same comparison. For example, 1:3, 2:6, and 3:9 are equivalent ratios.

• To find an equivalent ratio, multiply both terms in a ratio by the same number. This number is called the **scale factor**. For example, 1:3 = (1 × 2):(3 × 2) = 2:6. The scale factor is 2.

• The number sentence 1:3 = 2:6 is an example of a **proportion**, a number sentence that shows two equivalent ratios.

3. Shade the grid on the right so that the ratio of shaded squares to total squares is the same as the ratio in the grid on the left. Then write the proportion.

   a)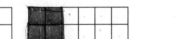

   b)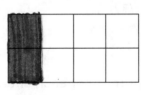

   d) 100:30
      50:15
      20:6

      C

   Proportion: ____8:16____

   Proportion: ____2:6____

4. In a mixture of garden soil, 2 pails of peat moss are mixed with every 3 pails of earth.

   a) Write the ratio of peat moss to earth. ____2:3____ $2\frac{5}{10}$

   b) If you increase the amount of peat moss to 10 pails, how many pails of earth will you need? $\frac{2}{3} : \frac{10}{15}$ The ratio is: 10:15 ✓

5. Paul is 12 years old. His brother Josh is 6 years old. Their father is six times as old as Josh.

   a) Write a ratio to compare Paul's age to Josh's age. ____12:6____

   b) How old is Paul's father? 6×6=36
      Paul's father is 36 years old.

      a   36:12

   c) How many times older than Paul is his father?
      $\frac{36}{12}$ = 24   36:12 = 3 times

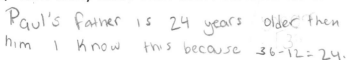
Paul's father is 24 years older then him I know this because 36-12=24.

2.3 Solving Rate Problems

▶ **GOAL:** Determining equivalent rates to solve rate problems.

1. Write each comparison as a rate. Then write two equivalent rates for each comparison.

    a) Romona went snowboarding nine times in three months.

    b) Miguel earns $32 dollars in 4 h.

2. Bonnie can ride her bike 70 km in 5 h.

    a) What is her average rate per hour?

    b) How long does it take Bonnie to cycle to her cousin's house, which is 28 km away?

3. Write a proportion for each situation. Determine the missing term in each proportion.

    a) Two shirts cost $36. One shirt costs $_____.

    b) You can earn $9/h. In 12 h, you can earn $_____.

4. Fill out the chart to find the number of heartbeats in 1 min for each.

	Animal	Heartbeats	Time (s)	Heartbeats/min
a)	elephant	9	15	
b)	human	6	5	
c)	cat	13	6	

5. Use the information in question 4 to answer these questions. How long does it take

    a) an elephant's heart to beat 180 times?

    b) a cat's heart to beat 520 times?

Communicating about Ratio and Rate Problems

▶ **GOAL: Explain your thinking when solving ratio and rate problems.**

Use the Communication Checklist to help you answer each question.

1. The Tigers have won 12 out of 15 hockey games. If they play 5 more games, would you expect them to win 4? Explain your thinking.

2. Fawn walks 2 km in 30 min.

   a) How far can she walk in 1 h? Explain your thinking.

   b) Would you expect Fawn to walk 9 km in 2 h? Explain your thinking.

3. Paul wants to earn $10 selling brownies at a bake sale. The brownies will be priced at two for $1.

   a) Paul thinks, "I know I earn $1 for every two brownies, and I want to know how many brownies I need to sell to earn $10." Complete Paul's explanation, and solve the problem.

   b) Paul expects to sell four out of every five brownies he bakes. He thinks, "I will only sell four out of five brownies, so I need to bake more brownies than I want to sell. I want to earn $10 in total. I need to know how many brownies to bake." Complete Paul's explanation, and solve the problem.

2.5 Ratios as Percents

▶ **GOAL:** Solve problems that involve conversions between percents, fractions, and decimals.

1. Each diagram is made from 100 squares. Write a fraction, a decimal, and a percent to show what part has been shaded.

a)

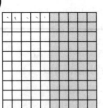

Fraction: $\frac{5}{10}$
Decimal: 0.5
Percent: 50%

c)

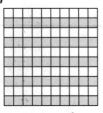

Fraction: $\frac{5}{10}$
Decimal: 0.5
Percent: 50%

e)

Fraction: $\frac{28}{100}$
Decimal: 0.28
Percent: 28%

b)

Fraction: $\frac{75}{100}$
Decimal: 0.75
Percent: 75%

d)

Fraction: $\frac{5}{10}$
Decimal: 0.5
Percent: 50%

f)

Fraction: $\frac{40}{100}$
Decimal: 0.40
Percent: 40%

$100 \div 2 = 50$

At-Home Help

A percent is used to compare a part to a whole. For example, you could say that 15% of all pets in Canada are hamsters. This means that 15 out of every 100 pets in Canada are hamsters.

Fifteen percent (15%) can also be written as a ratio (15 : 100), a fraction $\left(\frac{15}{100}\right)$, or a decimal, 0.15.

To convert a ratio to a percent, find an equivalent ratio out of 100. For example, to convert the ratio 2 out of 5 to a percent, you could write the following:

2 : 5 = ▨ : 100

▨ = 40, or 40%

2. Write each fraction as a percent.

   a) $\frac{46}{100} = $ _____46_____ %

   b) $\frac{19}{50} = $ _____ %

   c) $\frac{12}{25} = $ _____ %

   d) $\frac{4}{10} = $ _____4_____ %

   e) $\frac{4}{5} = $ _____ %

   f) $\frac{3}{4} = $ _____ %

3. You are given four percents: 21%, 74%, 98%, and 3%.

   a) Write each percent as a decimal.

   21% = _____

   74% = _____

   98% = _____

   3% = _____

   b) Write each percent as a ratio.

   21% = _____

   74% = _____

   98% = _____

   3% = _____

4. Fill in the missing equivalent fraction, ratio, decimal, or percent in each row.

	Fraction	Ratio	Decimal	Percent
a)	$\frac{1}{4}$	1:4	0.25	25%
b)	$\frac{3}{5}$	3:5	0.6	60%
c)	$\frac{1}{5}$	1:5	.5	20%
d)	$\frac{3}{10}$	3:10	0.30	30%
e)	$\frac{41}{100}$	41:100	0.	41%
f)	$\frac{9}{50}$	9:50	0.18	18%
g)	$\frac{23}{}$	2:3	0.6	23%
h)	3:1	6:8	0.75	6%

5. Arrange in order from greatest to least.

   a) 41%, 64%, 100%, 2%, 84%: _____

   b) 7%, 14%, 114%, 35%, 3%: _____

   c) 19%, 0.5, 74%, 0.32, 0.9: _____

   d) $\frac{45}{100}$, 88%, 56%, $\frac{2}{100}$: _____

   e) $\frac{1}{4}$, 0.81, 85%, $\frac{2}{10}$, 44%: _____

   f) 72%, 8/20, 0.04, 91%, $\frac{26}{100}$: _____

6. Kwami ate 3 crackers from a box of 12. What percent of the crackers did he eat?

7. Paul earned $20 by selling brownies. He needs $100 in total. What percent of the money has Paul earned?

8. At a restaurant, Jody tipped the waiter $3. The cost of the meal was $20. What percent of the cost of the meal did Jody give to the waiter?

2.6 Solving Percent Problems

▶ GOAL: Solve percent problems using equivalent ratios and decimals.

1. Calculate.

   a) 30% of 100 = _____

   b) 20% of 50 = _____

   c) 15% of 40 = _____

   d) 12% of 30 = _____

   e) 75% of 90 = _____

   f) 15% of 80 = _____

At-Home Help

Here are some methods you can use to solve percent problems.
• Write a proportion, and solve to find the missing term.
• Use logical reasoning to do a mental calculation.
• If you need to find a percent of a number, express the percent as a decimal, and multiply the number by it.

2. Find the missing number.

   a) 10% of _____ = 30

   b) 40% of _____ = 60

   c) 50% of _____ = 35

   d) 20% of _____ = 150

   e) 25% of _____ = 200

   f) 75% of _____ = 30

3. Water covers 72% of Earth's surface. What percent is dry land? __28%__

   $100 - 72 = 28\%$

   $$\begin{array}{r} \overset{0\ \ \overset{9}{\cancel{10}}\ \overset{1}{\cancel{0}}}{\cancel{100}} \\ -\ \ 72 \\ \hline 28 \end{array}$$

4. The ratio of the number of people at a dinner to the number invited was 106 to 100. Express this as a percent.

   $$\begin{array}{r} \overset{1}{}\overset{3}{}\overset{9}{}\overset{1}{} \\ \$4.00 \\ -\ \ 6.75 \\ \hline 7.25 \end{array} \qquad \begin{array}{r} 6.75 \\ +\ 7.25 \\ \hline 14.90 \end{array}$$

5. Miguel is comparing the sale price of his favourite CD at three stores. At all stores the regular price of the CD is $24. Able Audio has a 33% off sale, Super Sound has a 25% off sale, and Mighty Music is taking $6.75 off the regular price of all CDs.

   a) Calculate the sale price of the CD at each store.

   CD price = $24   Mighty music

   Able Audio = 33% $$\begin{array}{r} \overset{1}{}\overset{3}{}\overset{9}{}\overset{1}{} \\ \$24.00 \\ -\ \ \$6.75 \\ \hline \$7.25 \end{array}$$

   Super Sound = 25%

   Mighty music = −$6.75

   b) Which store has the lowest sale price? _____

6. Fawn got her results back from three tests. The first test was marked out of 50, the second test out of 80, and the third test out of 20. Her percent on each test was the same: 80%. How many marks did Fawn get on each test?

7. According to an advertisement, 45% of the people in Sunnydale eat Baker's Bread. If 1300 people live in Sunnydale, how many of them eat Baker's Bread?

8. In a badminton tournament, Ravi and Chang won 75% of their matches.

   a) If there were 16 matches in total, how many did Ravi and Chang win?

   b) How many did they lose? _____

9. There are 950 students at Sunnydale Elementary School. On School Spirit Day, 285 of the students wore the school colours, red and blue. Express this number as a percent.

10. At a clothing store, all the shirts are on sale. The following chart lists the items that are on sale. Fill out the chart to calculate the new sale prices.

	Item	Price	Discount	Amount off	Sale price
a)	white shirt	$30.00	10%		
b)	red shirt	$15.00	33%		
c)	blue shirt	$45.00	50%		
d)	green shirt	$50.00	5%		
e)	yellow shirt	$98.00	15%		
f)	black shirt	$22.00	75%		

11. Miguel bought a bicycle at a 30% discount. He paid $122.50.

   a) What percent of the original price did Miguel pay? _____

   b) What was the original price?

   c) How much money did Miguel save?

2.7 Decimal Multiplication

▶ **GOAL: Use decimal multiplication to solve ratio and rate problems.**

1. Use the 10-by-10 grids to model, and then calculate.

   a) 0.3 × 0.5 = _____

   b) 0.6 × 0.9 = _____

   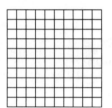

At-Home Help

The following strategies can help you multiply decimals:
- Use a 10-by-10 grid to model the problem.
- Multiply one of the numbers by a power of 10 (for example, 10, 100, 1000) to obtain a whole number. Then multiply it with the other number. Divide your answer by 10, 100, or 1000 (whichever you used).
- Round the numbers to the nearest whole number, and multiply. Use your estimated answer to check whether your final answer is reasonable.

2. Estimate, and then calculate. Show the strategy you used.

   a) 0.7 × 0.2

   b) 0.4 × 1.2

   c) 3.1 × 0.5

   d) 0.06 × 1.6

3. Calculate the distance a cyclist will travel if she cycles at 16.5 km/h for 2.5 h.

4. Tynessa bought 4.2 m of ribbon that cost $0.75 per metre. How much money did she spend?

2.8 Decimal Division

▶ **GOAL: Use decimal division to solve ratio and rate problems.**

1. Use the 10-by-10 grids to model, and then calculate.

   a) $2.6 \div 0.2 =$ _____

   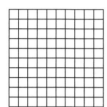 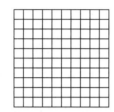 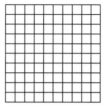

   b) $1.8 \div 0.15 =$ _____

   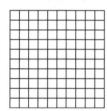 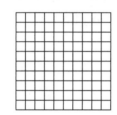

At–Home Help

The following strategies can help you divide decimals:
- Use 10-by-10 grids to model the problem.
- Write the numbers in the form of a fraction. Multiply both the top and bottom of the fraction by a power of 10 (for example, 10, 100, 1000) to obtain whole numbers. Then divide.
- Round the numbers to the nearest whole number, and divide. Use your estimated answer to check whether your final answer is reasonable.

2. Divide.

   a) $3.2 \div 0.8$ **c)** $3.12 \div 0.75$

   b) $8.25 \div 0.25$ **d)** $0.736 \div 0.08$

3. You are given 15.30 m of fabric for making flags. About how many flags can you make from the fabric if each flag is

   a) half a metre wide   **b)** 3 m wide   **c)** 90.5 cm wide

4. Romona earns $9.50/h at her job. She earned $118.75 last week. How many hours did she work?

Test Yourself

1. Look at the figure and answer the questions.

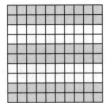

   a) What is the ratio of white squares to grey squares? _____

   b) What fraction of the squares is white? _____

   c) What percent of the squares is grey?

   _____

2. Which of the following are equivalent ratios?

   a) 6:8

   b) 9:15

   c) 12:16

   d) 15:20

3. Find the missing term.

   a) 3:6 = 2:▨

   b) 4:10 = ▨:15

   c) $\dfrac{2}{5} = \dfrac{6}{▨}$

   d) $\dfrac{16}{24} = \dfrac{▨}{6}$

   e) 5:15 = ▨:3

   f) 4:1 = 32:▨

   g) 14:7 = 6:▨

   h) 10:2 = ▨:6

4. If 8 people can eat 8 eggs in 8 min, how long will it take 7 people to eat 7 eggs?

5. Fill in the chart to find out which person earns the most money per hour.

	Hours (h)	Earnings ($)	Earnings per hour ($/h)
Miguel	32	384.00	
Bonnie	12	102.00	
Paul	25	250.00	
Romona	9	69.75	

6. At a football game, 30% of the people are waving flags. If there are 4500 people at the game, how many are waving flags?

7. Calculate.

   a) 50% of 220 = _____

   b) 25% of 40 = _____

   c) 15% of _____ = 9

   d) _____% of 35 = 7

   e) 20% of _____ = 28

   f) 4% of 550 = _____

8. The ratio of cats to dogs at a pet show is 1:3.

   a) What percent of the pets at the show are cats?

   b) If there are 25 cats, how many dogs are at the show?

9. Estimate, then calculate.

a) 1.2×4.4

Estimation: _____

Calculation: _____

b) 0.03×0.5

Estimation: _____

Calculation: _____

c) 3.3×0.6

Estimation: _____

Calculation: _____

d) $3.2 \div 0.8$

Estimation: _____

Calculation: _____

e) $4.5 \div 0.9$

Estimation: _____

Calculation: _____

f) $5.22 \div 1.34$

Estimation: _____

Calculation: _____

10. Miguel, Fawn, Paul, and Romona received marks of 87%, $\frac{30}{40}$, 79%, and $\frac{24}{25}$, respectively. Express their marks as decimals. Then arrange their marks in order from greatest to least.

11. At the corn roast, 3 cobs were cooked for every 7 persons. If 144 cobs were cooked, how many people were at the roast?

12. Indira's bedroom has been drawn to scale. In the drawing, 0.5 cm = 30 cm. Each square measures 0.5 cm by 0.5 cm.

Calculate the following:

a) the length and width of Indira's bed:

_____

b) the width of the bedroom window:

_____

c) the dimensions of the rug:

_____

d) the length of the bookcase:

_____

Collecting Data

▶ **GOAL:** Make inferences and convincing arguments that are based on primary and secondary data.

Read the story, then answer the questions that follow.

The Grade 7 classes at Sunnydale Elementary School were planning a big year-end picnic. Each student was asked to complete the following survey:

Picnic Survey	
Meal choices	Which would you prefer as your main meal? a) pizza c) hamburgers b) hot dogs d) sandwiches
Drink choices	Which type of juice would you prefer as your beverage? a) apple c) grape b) orange

At-Home Help

- **Primary data** is information that is collected directly.
- **Secondary data** is information that is collected by someone else.

Out of 180 students, 27 chose pizza, 90 chose hot dogs, 54 chose hamburgers, and 9 chose sandwiches as their main meal. For their drink, 126 students chose apple juice, 45 chose grape juice, and 9 chose orange juice.

1. Are the survey results primary or secondary data? Explain your answer.

   _____

   _____

2. **a)** What percent of students chose each type of food for their main meal?

   b) What two foods would you serve at the picnic? Use the data to back up your choice.

   _____

   _____

3. **a)** What percent of students chose each type of drink?

   b) Which drink would you serve at the picnic? Use the data to back up your choice.

   _____

Avoiding Bias in Data Collection

▶ **GOAL:** Understand different ways to collect data and analyze bias in data-collection methods.

1. In each of these situations, circle which method you would choose to collect data.

   a) To determine where students at your school eat their lunch,

      i) ask the first 10 people in the cafeteria line.

      ii) ask every 3rd person in your math class.

      iii) ask the first 10 people going home for lunch.

   b) To determine what students at your school think about starting a lunch-hour basketball tournament,

      i) ask every 3rd student that enters the school library before classes, at lunch, and after classes.

      ii) ask every 10th student entering the school.

      iii) ask every teacher in the school.

   c) To determine what percent of preschoolers stay home with a parent or caregiver and what percent go to daycare,

      i) interview every 10th household with a child between the ages of 0 and 5, as determined by the Canadian census.

      ii) call every 10th number in the phone book during the day, and ask whoever answers if he or she has children at home or in daycare.

      iii) ask all Grade 7 students how many preschoolers they know who stay home with a parent or caregiver, and how many they know who go to daycare.

2. Listed below are poorly worded survey questions. The results of these surveys will be biased. Write new questions that correct this problem.

   a) Kwami was carrying out a survey on favourite breakfast cereals. He asked, "Do you like cereal?"

   b) Indira wants to find out what type of music people like to listen to. She asks, "Do you like pop music or classical music?"

At-Home Help

When the results of a survey of one group are not likely to apply to another group selected from the same population, the results are **biased**. For example, suppose you want to find out how many Canadians like hockey. You would get biased results if you surveyed only the members of a seniors' knitting club or only the people at a hockey game. To get unbiased results, you need to survey a group of people that can represent the whole population. For example, you could conduct a survey to ask 1 out of every 10 people in your neighbourhood.

Using a Database

▶ **GOAL:** Use a database to sort and locate data.

The clothing store where Indira works uses a database to organize information about items in the store. Use the information in the database at the bottom of the page to answer the questions.

At-Home Help

- A **database** is an organized set of information, often stored on a computer.
- A **record** is all the data about one item in the database.
- A **field** is a category used as part of a database.
- An **entry** is a single piece of data in a database.
- To **sort** is to order information from greatest (or first) to least (or last). A database can be sorted by fields.

1. **a)** Is the column titled "Colour" a record, a field, or an entry? Explain your answer.

   b) Is the shaded row a record, a field, or an entry? Explain your answer.

2. What field has been used to sort the items? _____

3. Indira wants to find the least expensive item currently on sale.

   a) What field should she use to sort the database? _____

   b) What is the least expensive item? _____

   c) What is the most expensive item? _____

4. The database is usually sorted by item code. What item would come fourth if the database were sorted in this way?

   _____

5. If the red dress appeared last, which field would Indira have used to sort?

Item code	Description	Colour	Original price	Sale price	Number in store
101556	sweater	yellow	$65.00	$55.00	5
101558	dress	red	$99.00	$69.00	7
101559	skirt	orange	$44.00	$29.00	8
101555	shirt	green	$39.00	$30.00	11
101557	sweater	black	$84.00	$75.00	12
101560	skirt	brown	$74.00	$49.00	13
101554	shirt	blue	$29.00	$25.00	15
101561	jeans	indigo	$25.00	$20.00	22

100

3.4 Using a Spreadsheet

▶ **GOAL:** Use a spreadsheet, and understand the difference between a spreadsheet and a database.

The spreadsheet at the bottom of the page represents a partial inventory at a sports store. Use the spreadsheet to answer the questions.

1. a) What is the price of running shoes at this store?

   _____

   b) How many sweat shirts does the store have?

   _____

2. a) What is shown in cell B4? _____

   b) What is shown in column C? _____

   c) What is shown in row 4? _____

3. The number 9.95 in cell B2 and the number 20 in cell C2 were entered directly. The formula B2*C2 was entered for cell D2. What calculation will the computer make for cell D2?

4. What formulas would you need to enter for cells D3 and D4 to calculate these two total costs?

5. What formula could you enter for cell D6 if you wanted the sum of all four items?

	A	B	C	D
1	Item	Cost per item	Number	Total cost
2	T-shirt	$9.95	20	$199.00
3	sweat shirts	$29.95	25	
4	shorts	$12.95	15	
5	running shoes	$49.99	12	
6				

3.5 Frequency Tables and Stem-and-Leaf Plots

▶ **GOAL: Organize data using frequency tables and stem-and-leaf plots.**

1. A survey was carried out to find the months in which people were born. The results are given in the frequency table below.

Birthday Months	
Interval (d)	**Frequency**
January 1–31	7
February 1–29	5
March 1–31	2
April 1–30	2
May 1–31	2
June 1–30	2
July 1–31	2
August 1–31	1
September 1–30	3
October 1–31	5
November 1–30	2
December 1–31	2

a) How many people were surveyed? _____

b) During which month were most people born? _____

2. The percent students received on their mathematics examination are given in the stem-and-leaf plot below.

Examination Mark	
Stem	**Leaf**
5	1 4 7
6	3 3 8 9
7	1 1 2 4 9
8	0 1 1 2 3 4 4 5 8
9	1 2 2 4 9 9

a) What was the highest mark on the examination? _____

b) What was the lowest mark on the examination? _____

c) How many people received a mark of 75% or over? _____

d) How many people received a mark higher than 75% but lower than 89%? _____

3. Use the chart below to create a frequency table to display the information in the stem-and-leaf plot in question 2.

Examination Mark	
Interval	Frequency

4. Bonnie wants to know how long it takes students to get ready for school in the morning. She asks all the students in her class to time themselves one morning, and obtains the following data:

Amount of Time to Get Ready for School in the Morning (min)							
66	22	75	38	49	96	19	76
45	16	40	73	52	44	28	48
31	50	35	93	67	74	64	50
25	62	14	55	88	102	81	33

a) Use the chart below to display Bonnie's data in a stem-and-leaf plot.

Amount of Time to Get Ready for School (min)	
Stem	Leaf

b) How many students are in Bonnie's class? _____

c) What is the shortest time to get ready for school? _____

d) What percent of the students in Bonnie's class take less than 20 min to get ready for school in the morning? Round off to the nearest percent.

3.6 Mean, Median, and Mode

▶ **GOAL:** Describe data using the mean, median, and mode.

1. Calculate the mean, median, and mode for these sets of data.

At-Home Help

• A **mean** is the sum of a set of numbers divided by the number of numbers in the set.
• A **median** is the middle value in a set of ordered data. When there is an even number of numbers, the median is the mean of the two middle numbers.
• A **mode** is the number that occurs most often in a set of data. There can be more than one mode, or there might be no mode.

a) 4, 6, 9, 8, 7, 9, 10, 4, 7, 6, 9, 9

Mean: _____ Median: _____
Mode: _____

b) 10, 15, 11, 17, 14, 16, 20, 13, 12, 15

Mean: _____ Median: _____
Mode: _____

c) 51, 56, 52, 58, 59, 54, 52, 57, 54, 59, 54

Mean: _____ Median: _____ Mode: _____

d) 3, 2, 2, 5, 9, 4, 8, 4, 3, 4

Mean: _____ Median: _____ Mode: _____

e) 23, 36, 43, 26, 76, 23, 23

Mean: _____ Median: _____ Mode: _____

f) 87, 58, 80, 76, 45, 2, 99, 2, 34, 56, 77, 5, 42, 96, 13, 56

Mean: _____ Median: _____ Mode: _____

2. This table gives the weekly wages earned by workers in a small factory.

Worker	Number of workers	Weekly wage
Machine operator	11	$750
Office worker	5	$800
Supervisor	3	$900
Production manager	1	$1300
Factory owner	1	$2500
Total	21	

a) For the weekly wage, calculate the mean, median, and mode.

Mean: _____ Median: _____ Mode: _____

b) If the factory owner was trying to hire employees, would he or she use the mean, median, or mode to advertise the weekly wage?

_____

c) If the machine operators wanted to ask for more money, would they use the mean, median, or mode to measure the weekly wage?

_____

3. The following table shows the number of juice boxes sold at the school cafeteria over the last week.

Type of juice	Tally	Frequency
Apple	ⅲ ⅲ II	
Orange	ⅲ ⅲ ⅲ ⅲ ⅲ ⅲ II	
Lemonade	ⅲ ⅲ ⅲ	
Grape	ⅲ ⅲ ⅲ ⅲ III	
Grapefruit	IIII	

a) Complete the frequency table.

b) What is the mean, median, and mode of all the juice boxes sold?

Mean: _____ Median: _____ Mode: _____

c) Based on this chart, which type of juice would you recommend that the cafeteria buy in larger quantities?

_____

d) Based on this chart, which juice would you recommend that the cafeteria stop selling?

_____

Communicating about Graphs

▶ **GOAL:** Make inferences and convincing arguments that are based on analyzing data and on trends.

1. Fifty people were asked to name their favourite flavour of potato chips. The results are shown in the bar graph.

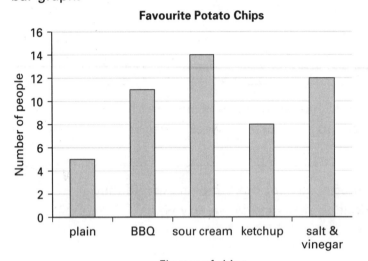

Favourite Potato Chips

a) Which is the favourite flavour? _____

b) Which is the least favourite? _____

c) What three flavours would you recommend that a cafeteria buy? Why?

_____

2. The graph below gives information about foods bought at a school cafeteria over the last six months.

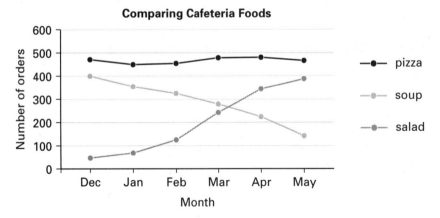

How can the cafeteria use this data to order food for the next month?

_____

Test Yourself

1. Tynessa wants to find out what percent of people in the neighbourhood use the local library. She could
   - survey 1 out of every 10 families in the neighbourhood
   - survey all the members of the neighbourhood book club
   - survey 1 out of every 15 students at her school

   a) Which group(s) will produce a bias in favour of using the library?

   _____

   _____

   _____

   b) Which group do you think Tynessa should survey? Why do you think so?

   _____

   _____

   _____

   _____

2. Simon's father runs a catering business. He stores information about the foods he prepares in the database at the bottom of the page.

   a) Currently, the foods are sorted alphabetically by the field "Name of dish." Identify the food that would come first if the foods were sorted by price, from lowest to highest:

   _____

   b) From the database, give one example of a numeric field:

   _____

3. Sandra is conducting a survey on TV viewing in her neighbourhood. Write an example of a question she could ask.

   _____

   _____

   _____

   _____

Name of dish	Number of people it serves	Main ingredients	Price
BBQ ribs	2	beef, barbecue sauce	$32.00
Chicken pot pie	6	chicken, potatoes, pastry	$25.00
Teriyaki stir-fry	3	tofu, snow peas, broccoli, bean sprouts	$19.00
Vegetable curry	4	broccoli, sweet peppers, potatoes, coconut milk	$22.00

100

4. Simon's father also uses a spreadsheet to keep track of orders over the weekend. Use the spreadsheet below to answer the questions.

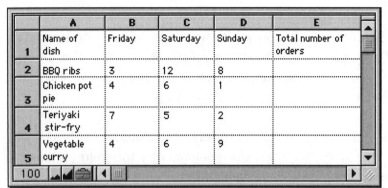

	A	B	C	D	E
1	Name of dish	Friday	Saturday	Sunday	Total number of orders
2	BBQ ribs	3	12	8	
3	Chicken pot pie	4	6	1	
4	Teriyaki stir-fry	7	5	2	
5	Vegetable curry	4	6	9	

a) What is the entry in cell C3?

_____

b) What formula should be in cell E2?

_____

c) What number should appear in cell E4?

_____

5. The track-and-field team made the following times on the 100 m race:

55   43   39   56   46   55   58   62

46   61   48   46   50   43   44

a) In the space below, draw a stem-and-leaf plot to record this data.

b) Calculate the mean, median, and mode for this data.

Mean: _____

Median: _____

Mode: _____

6. Chang surveyed the students in his class about how they spend their time each day. He displayed his data in the following circle graph:

Students' Time in Hours

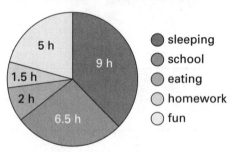

- sleeping
- school
- eating
- homework
- fun

5 h
9 h
1.5 h
2 h
6.5 h

a) If each number on the graph represents a number of hours, how many hours do the students in Chang's class spend sleeping?

_____

b) What percent of the day do the students spend doing homework?

Exploring Number Patterns

▶ **GOAL: Identify and describe number patterns.**

1. Follow the pattern rule on the left to create a pattern of numbers starting and ending with the numbers on the right.

	Pattern rule	Pattern of numbers
a)	Subtract 2 from each number to get the next number in the pattern.	36, ____, ____, ____, ____, ____, 24
b)	Add 10 to each number to get the next number in the pattern.	6, ____, ____, ____, ____, ____, 66
c)	Multiply each number by 2 to get the next number in the pattern.	3, ____, ____, ____, ____, ____, 192
d)	Divide each number by 10 to get the next number in the pattern.	1000, ____, ____, ____, ____, ____, 0.001

2. Using the two numbers above each box in part (b),

a) determine the pattern rule.

_____

b) complete the pattern by filling in the boxes.

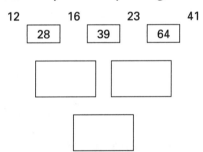

3. Examine each of the following patterns, and then draw the next two items in the pattern.

a)

b)

Applying Pattern Rules

▶ **GOAL:** Recognize patterns, and use rules to extend and create the patterns.

1. Describe the pattern rule for each sequence. Write the next three numbers.

   a) 0, 3, 6, 9, 12, _____, _____, _____

   Pattern Rule: _____

   b) 1, 5, 9, 13, 17, _____, _____, _____

   Pattern Rule: _____

   c) 1, 4, 16, 64, _____, _____, _____

   Pattern Rule: _____

   d) 100, 91, 82, 73, _____, _____, _____

   Pattern Rule: _____

   e) 5, 10, 20, 40, 80, _____, _____, _____

   Pattern Rule: _____

   f) 1280, 640, 320, 160, _____, _____, _____

   Pattern Rule: _____

   g) 2, 12, 72, 432, _____, _____, _____

   Pattern Rule: _____

   h) 2, 3, 5, 9, 17, 33 _____, _____, _____

   Pattern Rule: _____

2. The pattern rule for a sequence is "Start with 1, and multiply each number by 5." Write the next five numbers in the sequence.

3. The pattern rule for a sequence is "Start with 4, multiply by 2, and subtract 1." Write the next five numbers in the sequence.

4. Describe the pattern rule for each sequence. Write the next three numbers.

   a) 0, 5, 10, 15, 20, _____, _____, _____

   Pattern Rule: _____

   b) 100 000, 10 000, 1000, _____, _____, _____

   Pattern Rule: _____

   c) $\frac{1}{1000}, \frac{1}{100}, \frac{1}{10}$, _____, _____, _____

   Pattern Rule: _____

   d) 1, 0.5, 0.25, 0.125, _____, _____, _____

   Pattern Rule: _____

   e) 1, 5, 25, 125, _____, _____, _____

   Pattern Rule: _____

   f) 333, 322, 311, 300, _____, _____, _____

   Pattern Rule: _____

   g) 0.2, 0.6, 1, 1.4, _____, _____, _____

   Pattern Rule: _____

   h) 2187, 729, 243, 81, _____, _____, _____

   Pattern Rule: _____

5. The pattern rule for a sequence is "Start with 2, double each number, and subtract 1." Write the first eight numbers in the sequence.

6. The pattern rule for a sequence is "Start with 1, double each number, and subtract 1." What is the 250th number in the sequence? How do you know?

7. a) Describe the pattern rule for this sequence.

   1, 1, 2, 6, 24, 120, ...

   Pattern Rule: _____

   b) Write the next three numbers.

   _____

4.3 Using a Table of Values to Represent a Sequence

▶ **GOAL:** Use tables of values to represent number sequences.

1. a) Complete the table of values for the pattern shown.

Term number	Picture	Term value
1		5
2		8
3		11
4		14
5		
6		

At–Home Help

The **term number** tells the position of a term in a sequence. For example, in the sequence 1, 3, 5, 7, 9, ..., the number 5 has the term number 3 because it is third in the sequence.

The **term value** is the numerical value of the term. For example, in the sequence 1, 3, 5, 7, 9, ..., term number 1 has a value of 1, term number 2 has a value of 3, term number 3 has a value of 5, and so on.

b) Write a pattern rule that tells how the value of each term can be calculated from the previous term in the sequence.

_____

_____

c) Write a pattern rule that tells how the value of each term can be calculated from its term number.

_____

_____

d) Predict the value of the eighth term in the sequence.

2. The terms in a sequence are given in the table below.

Term number	1	2	3	4	5	6	7	8
Term value	7	14	21	28				

a) Fill in the table for terms 5 to 8.

b) Write a pattern rule that tells how the value of each term can be calculated from its term number.

_____

_____

c) What is the value of the 20th term? _____

3. The terms in a sequence are given in the table below.

Term number	1	2	3	4	5	6	7	8
Term value	6	11	16	21				

a) Fill in the table for terms 5 to 8.

b) Write a pattern rule that tells how the value of each term can be calculated from its term number.

_____

_____

c) What is the value of the 16th term? _____

4. You are given a pattern rule to build a pattern out of toothpicks. The rule is "Start with a triangle, and add two toothpicks each time to make another triangle."

a) Use toothpicks or draw pictures showing the first four terms in the sequence.

b) Fill out this table of values to record the number of toothpicks you used to build each term in part (a).

Term number	1	2	3	4
Term value				

c) Write a rule that describes the pattern.

_____

d) Predict how many toothpicks you would need to build the 10th term in the sequence.

Solve Problems Using a Table of Values

▶ **GOAL:** Use a table of values to solve patterning problems.

1. Fill out the table of values below. Then predict the number of boxes there would be in the 15th figure.

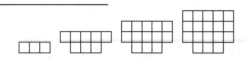

Term number	1	2	3	4	5	6	7	8
Term value (number of boxes)								

2. For Valentine's Day, each person in Sandra's study group gave a Valentine card to each other person.

a) If there were only one person (Sandra) in her group, there would be zero cards. If there were two people in her group, how many cards would there be? _____

b) Sketch what would happen if there were three people in Sandra's group. How many cards would be given out? _____

c) Sketch what would happen if there were four people in Sandra's group. How many cards would be given out? _____

d) Use the four values you know (one person, two people, three people, and four people) to start the table of values below.

Number of people	1	2	3	4	5	6	7
Number of cards							

e) What is the pattern rule?

_____

f) There are seven people in Sandra's group. How many Valentine cards were given out?

3. There are 10 soccer teams in the Sunnydale soccer league. Each team will play every other team once. How many games will there be in total?

4. James stacks boxes at a grocery store as a part-time job. On his first day, he stacked 16 boxes. Each day after that, he stacked 5 more boxes than the day before. On which day of work did James stack 46 boxes?

5. Yuki's guppies have begun to have babies. She started with 5 guppies. One week later, Yuki counted twice as many guppies in the fish tank. Each week afterwards, the number of guppies doubled. In how many weeks did Yuki have 160 guppies?

6. To raise money for a class trip, Rana's class sells granola bars. On the first day, 20 bars are sold. On the following days, the number of granola bars sold increases by 5 bars per day. Rana's class wants to raise $650.

   a) How many granola bars in total are sold by the end of the 6th day?

   b) Each granola bar costs $1. On which day of sales will the students reach their goal?

7. Omar wants to clear out the bushes in his backyard so he can flood it to make an ice rink in the winter. He counted 140 small bushes. On the first day, Omar cleared 7 bushes. Each day after that, he cleared 3 more bushes than he did the day before. How many days will it take for Omar to finish clearing bushes?

Using a Scatter Plot to Represent a Sequence

▶ **GOAL:** Use scatter plots to represent number sequences.

1. Create a scatter plot for the table of values. Use your scatter plot to find the missing values in the table.

Term number	Term value
1	
2	7
3	
4	15
5	
6	23

2. Mohammed is building a fence with two rails and one post in each section.

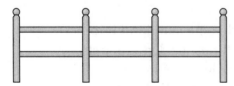

Fill in the missing values in the table on page 43. Then create a scatter plot on the grid to determine how many posts and rails Mohammed will need to build each fence.

a) A fence with 5 sections: _____

b) A fence with 11 sections: _____

Number of sections	Number of posts	Number of rails
0	1	0
1	2	2
2	3	
3		
4		
5		
6		

3. Colin is designing a series of paper chains. The second paper chain has 9 links. The third chain has 13 links, and the fourth chain has 17 links.

a) If all of Colin's chains follow the same sequence, how many links were there in Colin's first chain? _____

b) Fill out the table of values to show the number of links in each chain.

Term number (chain number)	Term value (number of links)

c) Create a scatter plot to show how many links there will be in Colin's ninth chain.

Test Yourself

1. Describe the pattern rule for each sequence. Write the next three terms.

    a) 2, 4, 6, 8, _____, _____, _____

    Pattern Rule: _____

    b) 0, 1, 3, 6, 10, _____, _____, _____

    Pattern Rule: _____

    c) 1, 4, 16, 64, _____, _____, _____

    Pattern Rule: _____

    d) 5, 8, 14, 26, _____, _____, _____

    Pattern Rule: _____

2. a) Fill out the blank table of values to predict how many circles you would need to build the seventh figure in this sequence.

Term number	Term value

    b) Draw the seventh figure.

    c) Explain how the pattern rule works.

    _____

3. Given the pattern rule, write the first five terms for each sequence.

    a) Start with 0, add 5.

    b) Start with 2, multiply by 6.

    c) Start with 100, divide by 2, add 10.

    d) Start with 1, multiply by 3, subtract 1.

4. Ravi is putting up a fence in his backyard. He builds 3 units the first day, 7 units the second day, and 11 units the third day. If he continues this pattern, how many days will it take him to finish the 55 unit fence? Complete the table of values to solve the problem.

Term number	Term value	Total units built
1	3	3
2	7	10
3	11	

5. a) Fill out the table of values for the numbers of white squares and shaded squares in each figure.

Figure number	Term value (white squares)	Term value (shaded squares)

b) Which figure in the sequence will have equal numbers of white and shaded squares?

_____

c) Draw a scatter plot on the grid to show how many white squares and how many shaded squares you would need to build the 10th figure.

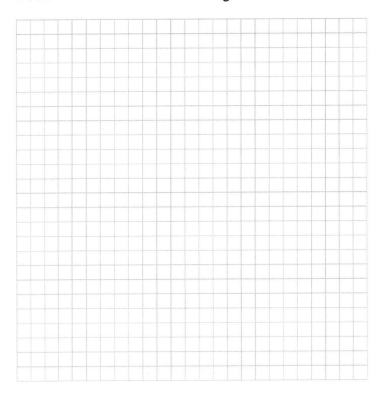

6. a) Make a scatter plot to represent this pattern of circles.

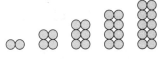

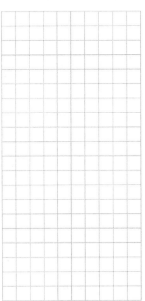

b) How many circles would you need to make the 8th figure in the sequence? _____

Area of a Parallelogram

▶ **GOAL: Apply the formula for the area of a parallelogram.**

1. For the following parallelogram:

   a) What is the height? _____

   b) What is the base? _____

   c) What is the area? _____

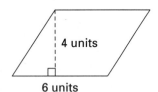

4 units

6 units

2. Fill in the blanks in the table.

	Base	Height	Area of parallelogram
a)	3 cm	5 cm	
b)	2 m		16 m²
c)		6 cm	30 cm²
d)	5.3 m	3.2 m	
e)		2.4 mm	3.6 mm²
f)	1.2 dm		0.6 dm²

3. Calculate the area of each parallelogram.

   A:

   B:

   C:

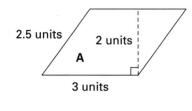

2.5 units   2 units

A

3 units

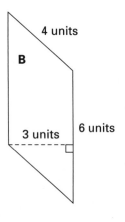

4 units

B

3 units   6 units

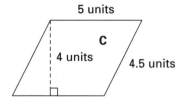

5 units

C

4 units   4.5 units

5.2 Area of a Triangle

▶ **GOAL: Apply the formula for the area of a triangle.**

1. Calculate the area of each triangle.

   a)

   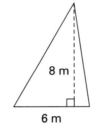

   8 m

   6 m

At–Home Help

You can think of a triangle as being half a parallelogram. The formula for the area of a triangle is the formula of a parallelogram divided in half:
$A = (b \times h) \div 2.$

   b)

   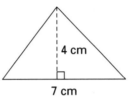

   4 cm

   7 cm

2. Fill in the blanks in the table.

	Base	Height	Area of triangle
a)	6 cm	12 cm	
b)		8 mm	32 mm^2
c)	120 m		1200 m^2
d)	14.2 cm	12.3 cm	

3. The base of a triangle is 30 cm. The height is 24 cm. Find its area.

4. Calculate the area of each shape by measuring the sides.

   a) $\triangle$ ABE   c) $\triangle$ BCD

   b) $\triangle$ BED   d) rectangle ACDE

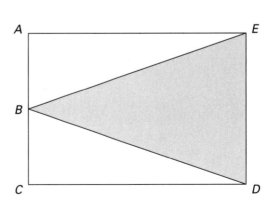

5.3 Calculating the Area of a Triangle

▶ **GOAL: Explore the area of a triangle.**

1. Triangle A, below, has an area of 3 units squared.

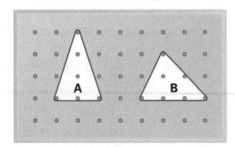

At-Home Help

The formula for the area of a triangle is the formula of a parallelogram divided in half:
$A = (b \times h) \div 2$.

a) Predict the area of triangle B. _____

b) Calculate the area of triangle B.

c) Triangle C has the same height as triangle A, but its base is double the base of triangle A. Predict the area of triangle C. Explain your answer.

_____

_____

2. Use the grid below to draw three different triangles that have an area of 24 units squared.

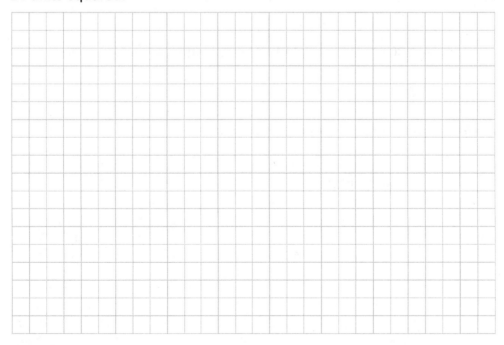

3. A triangle has a base of 4 m and a height of 10 m.

   a) What is the area of the triangle?

   b) A second triangle has a base of 4 m and a height of 8 m. Do you expect the area of this triangle to be greater than, less than, or equal to the area of the first triangle? Explain your answer.

   _____

   c) Calculate the area of the second triangle.

   d) A third triangle has a base of 10 m and a height of 4 m. Do you expect the area of this triangle to be greater than, less than, or equal to the area of the first triangle? Explain your answer.

   _____

   _____

4. Simon's bedroom window is 100 cm wide and 160 cm high. He wants to decorate the window with a large triangle of see-through light-green plastic.

   a) What is the largest area the triangle could have?

   b) Simon decides to use a triangle that has a base half as wide as the window, and a height half as tall as the window. What is the area of the triangle?

   c) On the grid below, draw two different triangles that Simon could use for his window decoration. Both triangles should have the dimensions given in part (b). Your triangles can be drawn to scale, so that 1 square = 10 cm.

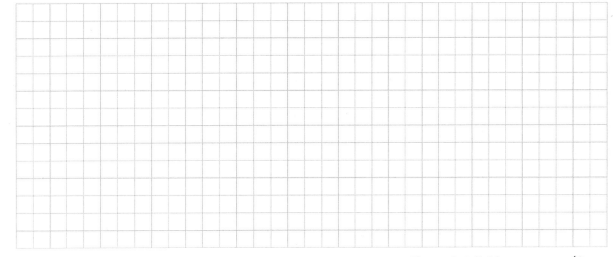

Area of a Trapezoid

▶ **GOAL: Apply the formula for the area of a trapezoid.**

1. Calculate the area of the trapezoids by dividing their shapes into simpler shapes.

   a) Area of trapezoid A

   b) Area of trapezoid B

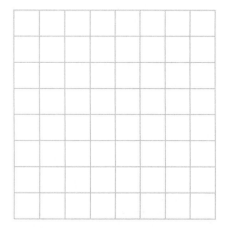

2. A trapezoid has parallel sides that are 16 cm and 12 cm long, and 20 cm apart. Calculate the area of the trapezoid.

3. A trapezoid has an area of 18 m², and parallel sides that measure 4 m and 2 m. What is the height of the trapezoid?

4. Use the grid to draw a trapezoid with an area of 9 units squared.

Exploring the Area and Perimeter of a Trapezoid

▶ **GOAL:** Explore the relationship between the area and the perimeter of a trapezoid.

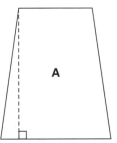

1. Measure the three trapezoids above, and fill out the table.

	Side length (cm)	Side length (cm)	Base a (cm)	Base b (cm)	Height h (cm)
Trapezoid A					
Trapezoid B					
Trapezoid C					

2. a) All three trapezoids have the same perimeter. What is it? _____

b) Predict which trapezoid has the greatest area. _____

3. a) Calculate the area for each trapezoid.

Trapezoid A: Trapezoid B: Trapezoid C:

b) Which trapezoid has the greatest area? _____

5.6 Calculating the Area of a Complex Shape

▶ **GOAL:** Calculate the area of an irregular 2-D shape by dividing it into simpler shapes.

1. Fill out the chart to find the total area of the polygon.

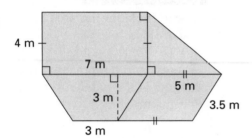

At-Home Help

To calculate the area of a complex shape, divide the shape into simpler shapes such as triangles, rectangles, parallelograms, and trapezoids. You can find the area of each simpler shape and add them together to find the total area.

Area of rectangle	Area of triangle	Area of parallelogram	Area of trapezoid

Total area of complex polygon: _____

2. Calculate the area of the shaded part of each diagram.

a)

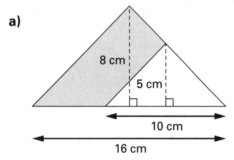

b)

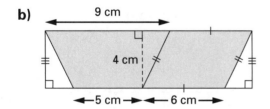

3. A park has a total area of 120 m². The park is divided into the following areas:

   - a rectangular playground with an area of 40 m²
   - a triangular picnic ground with an area of 38 m²
   - a field in the shape of a parallelogram

   a) What is the area of the field?

   b) The length of the field is 7 m. What is its height?

4. Mr. Dixon wants to carpet the first floor of his house. The first floor has the following rooms:

   - a square living room, 5 m by 5 m
   - a triangular kitchen, base of 4 m, height of 4 m
   - a rectangular hallway, 9 m by 1 m
   - an entranceway in the shape of a trapezoid, 2 m on one side, 3 m on the other side, and a height of 2 m

   a) How many m² of carpeting does Mr. Dixon need?

   b) If the carpet costs $8 per m², how much will the carpet cost in total?

5. Calculate the area of each polygon.

   a)

   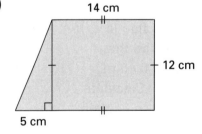

   b)

   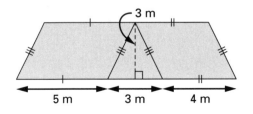

   c)

   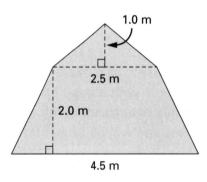

   d)

   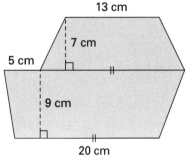

Communicating about Measurement

▶ **GOAL:** Describe perimeter and area using appropriate mathematical language.

1. Calculate the perimeter and area of the triangle. Explain what you did.

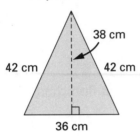

At-Home Help

Use this Communication Checklist to help you write clear and complete explanations.

Communication Checklist
☑ Did you include all the important details?
☑ Did you use correct units, symbols, and vocabulary?
☑ Did you completely solve the problem?

2. Calculate the area of the shaded part of each diagram. Explain what you did.

a)

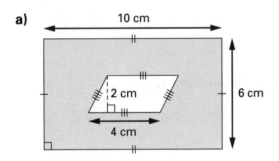

b)

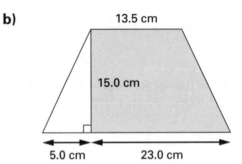

3. Sandra wants to make a flag out of red and blue fabric. The flag is a rectangle with a base of 1.0 m and a height of 0.50 m. Inside the rectangle is a triangle with a base of 0.50 m and a height of 0.40 m. The triangle will be blue, and the rest of the flag will be red. Calculate the area of the red part of the flag. Use words and a picture to explain what you did.

Test Yourself

1. Four parallelograms have the following bases and heights. Calculate the area of each.

   a) base: 5 cm, height: 3 cm

   b) base: 3.5 cm, height: 7 cm

   c) base: 6.3 cm, height: 2.2 cm

   d) base: 7.1 m, height: 3.0 m

2. A playground has two triangles painted on the ground. The first triangle has a base of 9 m and a height of 8 m. The base of the second triangle is one-third of the base of the first triangle, but its height is the same. Calculate the area of each triangle.

   First Triangle

   Second Triangle

3. Calculate the area of each figure.

   a)

   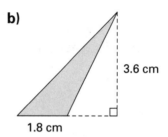

   2 m

   2 m

   b)

   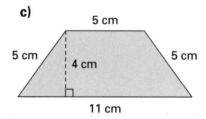

   3.6 cm

   1.8 cm

   c)

   5 cm

   5 cm   4 cm   5 cm

   11 cm

   d)

   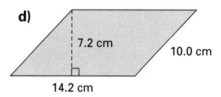

   7.2 cm

   10.0 cm

   14.2 cm

   e)

   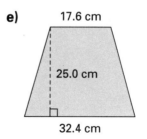

   17.6 cm

   25.0 cm

   32.4 cm

4. Calculate the area of the complex polygon below.

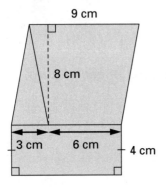

9 cm

8 cm

3 cm   6 cm   4 cm

5. Calculate the area of the grey shape in each diagram. Explain what you did.

a)

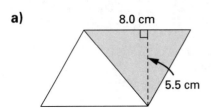

8.0 cm

5.5 cm

b)

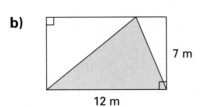

7 m

12 m

c)

5 cm   16 cm

10 cm

12 cm

6. Calculate the area of each polygon.

a)

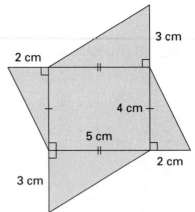

3 cm

2 cm

4 cm

5 cm

2 cm

3 cm

b)

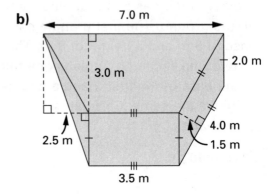

7.0 m

3.0 m

2.0 m

2.5 m

4.0 m

1.5 m

3.5 m

6.1 Comparing Positive and Negative Numbers

▶ **GOAL:** Compare and order positive and negative numbers.

1. Order these integers from least to greatest on the number line.

$-8, +8, -2, -5, +7, -1, 0, +3, -4, -7, +1, +5$

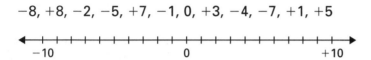

-10 0 $+10$

At-Home Help

Integers are all positive and negative whole numbers, including zero. It can help to think about integers in terms of temperatures. For example, $-5°C$ is warmer than $-15°C$ but colder than $+2°C$. Similarly, -5 is greater than -15 but less than $+2$.

2. Order these integers from least to greatest. Do not use a number line.

   a) $-4, +4, -3, +3, 0$: ____, ____, ____, ____, ____

   b) $-2, -6, +9, +5, -4$: ____, ____, ____, ____, ____

   c) $+22, -6, +35, -98, +1$: ____, ____, ____, ____, ____

   d) $-38, +45, -67, +8, 0$: ____, ____, ____, ____, ____

   e) $+3, -8, +46, +98, -123$: ____, ____, ____, ____, ____

3. Use the number line from question 1 to help you answer these questions.

   a) Which integer is 6 greater than -5? ____

   b) Which integer is 8 less than $+6$? ____

   c) Which integer is 1 more than -7? ____

   d) Which integer is halfway between $+4$ and -4? ____

   e) Which integer is halfway between -5 and $+3$? ____

   f) Which integer is 7 greater than -3? ____

   g) Which integers are between -2 and $+2$? ____, ____, ____

   h) Which integer is neither positive nor negative? ____

4. Enter $>$ or $<$ to compare the following pairs of integers.

   a) $(+3)$ ___ (-3) **d)** $(+8)$ ___ $(+9)$ **g)** (-46) ___ (0)

   b) (-8) ___ (-3) **e)** $(+7)$ ___ (-1) **h)** $(+3)$ ___ (-2367)

   c) (-7) ___ (-9) **f)** $(+27)$ ___ (-48)

6.2 An Integer Experiment

▶ **GOAL: Add positive and negative numbers.**

1. Ellie Vator works in the mailroom of a large corporation. She must deliver mail to all floors, from the lowest parking level (−5) to the penthouse (+35). Use the information to find out which floor she winds up on at the end of a long, tiring day!

   Ellie starts in the mailroom on floor 3. She rides the elevator up six floors (+6) to floor 9, then down three (−3) to floor 6. The rest of her deliveries are expressed as integers.

   Start at floor 6 (+6). Then,
   −8, −2, +22, −14, +30, −5, −14, +2, −7, +15, −9, −5, +12

   Where is Ellie?

   She is on floor _____ .

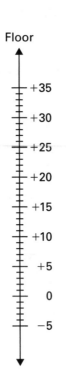

Floor

+35
+30
+25
+20
+15
+10
+5
0
−5

2. Do I understand the meaning of integers?

   To find the answer, shade the box that has the integer with the least value in each row. Then write the letter with that integer in the box on the right. The first one is done you.

O = +4	P = −4	Q = −2	R = +2	S = 0	P
L = −3	M = −13	N = −8	O = −18	P = −6	
O = +9	P = +3	Q = +5	R = +7	S = 0	
I = −33	J = +3	K = −28	L = +5	M = −22	
S = +9	T = +1	U = +88	V = +34	W = +5	
I = −64	J = −42	K = −9	L = −33	M = −2	
R = +3	S = −6	T = −12	U = +56	V = −24	
C = −6	D = +5	E = −14	F = −3	G = +52	
H = +4	I = +9	J = +409	K = +94	L = +1	
V = −43	W = +7	X = −83	Y = −92	Z = +3	

6.3 Adding Integers Using the Zero Principle

▶ **GOAL:** Use the zero principle, with and without models, to add integers.

1. Add these integers with the same signs.

a) $(+3) + (+4) =$ ____   c) $(+5) + (+3) =$ ____

b) $(-4) + (-2) =$ ____   d) $(-6) + (-5) =$ ____

2. Use the zero principle to add.

a) $(-1) + (+1) + (-1) + (+1) + (+1) =$ ___

b) $(+1) + (+1) + (-1) + (+1) =$ ____

3. Use opposite integers to add.

a) $(-5) + (+3) =$ ___   c) $(-10) + (+8) =$ ___

b) $(+7) + (-4) =$ ___   d) $(+15) + (-10) =$ ___

4. Write and solve an integer equation for the following statement: A stock opens at $4 then drops $1. What is it worth now?

5. Shade all boxes in which the sum is a positive integer.

(-3) $+(-8)$	(-4) $+(+2)$	(-5) $+(+6)$	$(+4)$ $+(-5)$	0 $+(-3)$	$(+5)$ $+(-8)$	(-2) $+(-3)$	(-4) $+\ 0$	$(+9)$ $+(-2)$	0 $+(-5)$
$(+8)$ $+(-9)$	(-5) $+(-6)$	(-3) $+(+8)$	0 $+(-2)$	$(+4)$ $+(-8)$	(-1) $+(-1)$	(-5) $+(-2)$	$(+5)$ $+(-2)$	(-6) $+(-9)$	(-6) $+\ 0$
$(+7)$ $+(-2)$	0 $+(+4)$	$(+5)$ $+(-3)$	(-3) $+(+7)$	$(+9)$ $+(+2)$	0 $+\ 0$	(-4) $+(+8)$	(-5) $+(-7)$	(-8) $+(-5)$	(-6) $+(-4)$
(-2) $+(+1)$	(-7) $+(-3)$	$(+4)$ $+0$	(-6) $+(-2)$	$(+3)$ $+(-7)$	$(+7)$ $+(-2)$	(-3) $+(-3)$	(-7) $+\ 0$	$(+2)$ $+(-8)$	$(+5)$ $+(-9)$
(-4) $+(-5)$	$(+7)$ $+(-8)$	$(+5)$ $+(+1)$	(-2) $+(-2)$	0 $+(+6)$	(-4) $+(-9)$	$(+8)$ $+(-7)$	(-4) $+(+5)$	0 $+(+9)$	(-2) $+(+8)$

6.4 Adding Integers That Are Far from Zero

▶ **GOAL: Add integers with and without models.**

1. Write the arithmetic statement for each model shown below. Write the final sum. The first one is done for you.

a) ⊖⊖⊖
+⊖⊖⊖
(−3) + (−3) = −

b) ⊖⊖⊖
+⊕⊕

c) ⊕⊕
+⊖⊖

d) ⊕⊕⊕
+⊖⊖⊖

e) ⊕⊕⊕
+ ⊖

f) ⊖⊖⊖⊖⊖
+⊕⊕⊕⊕⊕⊕

At-Home Help

When adding numbers that are not close to zero, use a number line to help you. Follow these three steps:

1. Draw an arrow from zero to the first number. For example, (−35) can be represented by an arrow 35 units long, starting at zero and pointing to the left.
2. Starting from the tip of the first arrow, draw a second arrow to represent the second number. For example, (+4) can be represented by an arrow 4 units long, pointing to the right.
3. The tip of the second arrow represents the answer. In this example, (−35) + (+4) = (−31).

2. Use the number lines to add these integers.

a) (−3) + (−5) = ____

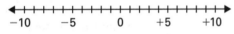

b) (+4) + (+3) = ____

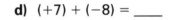

c) (−3) + (+4) = ____

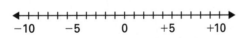

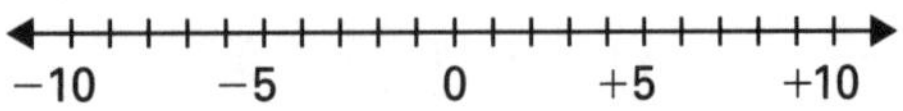

d) (+7) + (−8) = ____

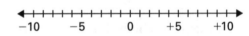

e) (−2) + (+9) = ____

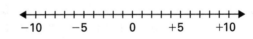

f) (−3) + (−2) + (−2) = _____

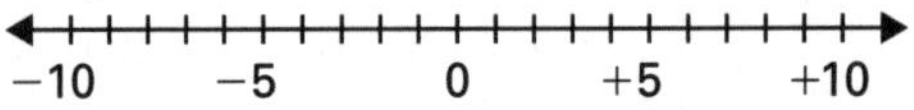

3. Add. Visualize a set of integer counters or a number line to help you.

a) (−45) + (+30) = ____

b) (−45) + (−30) = ____

c) (+45) + (+30) = ____

d) (+45) + (−30) = ____

e) (−100) + (−25) = ____

f) (−100) + (+25) = ____

g) (+100) + (+25) = ____

h) (+100) + (−25) = ____

i) (−68) + (+34) = ____

j) (−25) + (−40) + (+15) = ____

k) (+50) + (−30) + (−40) = ____

l) (+150) + (−75) + (−150) = ____

6.5 Integer Addition Strategies

▶ **GOAL: Learn integer addition strategies.**

1. Add these integers. First find pairs of opposite integers to simplify addition.

   a) $(-45) + (+48) + (+45) + (+6) =$ ____

   b) $(+23) + (+87) + (-23) + (+5) =$ ____

   c) $(+76) + (-69) + (-8) + (-76) =$ ____

   d) $(+6) + (-43) + (-6) + (-56) + (+43) =$ ____

   e) $(+53) + (-42) + (+9) + (+42) =$ ____

   f) $(-925) + (+387) + (+925) =$ ____

   g) $(-5) + (-3) + (-7) + (+7) =$ ____

2. Add these integers. Combine values to make pairs of opposite values to simplify addition.

   a) $(-3) + (-5) + (+8) + (-6) =$ ____

   b) $(+4) + (+3) + (-5) + (-7) =$ ____

   c) $(-30) + (+40) + (-10) + (-25) =$ ____

   d) $(+75) + (-80) + (+43) + (+5) =$ ____

   e) $(-200) + (+90) + (+110) + (-6) =$ ____

   f) $(-36) + (-26) + (-26) + (+52) =$ ____

3. Add these integers. Use regrouping to make more manageable sums.

   a) $(-450) + (+300) + (+80) =$ ____   **d)** $(-100) + (-25) + (+75) =$ ____

   b) $(-145) + (+100) + (-25) =$ ____   **e)** $(-100) + (+25) + (-7) =$ ____

   c) $(+24) + (-30) + (-82) =$ ____   **f)** $(+80) + (+25) + (-50) =$ ____

4. Add these integers. Use the strategy of your choice.

   a) $(-58) + (+40) + (+8) =$ ____

   b) $(+100) + (-57) + (+57) =$ ____

   c) $(+46) + (+34) + (-20) + (-26) =$ ____

   d) $(-125) + (-34) + (+125) + (-16) =$ ____

6.6 Using Counters to Subtract Integers

▶ **GOAL:** Subtract integers using a model.

1. Subtract these integers. You will not need to add zeros.

  a) $(-11) - (-7) = $ ____

  b) $(-42) - (-5) = $ ____

  c) $(+36) - (+30) = $ ____

2. Add zeros to subtract these integers.

  a) $(-7) - (-11) = $ ____ **d)** $(+20) - (-30) = $ ____

  b) $(+1) - (-6) = $ ____ **e)** $(+30) - (+36) = $ ____

  c) $(-24) - (-32) = $ ____ **f)** $0 - (-2) = $ ____

3. Draw a counter model for each difference. Calculate each difference.

  a) $(-5) - (+3) = $ _____ Model:

  b) $(-3) - (-2) = $ _____ Model:

  c) $(+4) - (+5) = $ _____ Model:

  d) $(+2) - (-4) = $ _____ Model:

At-Home Help

Using counters, such as white and black checkers, can help you subtract integers. For example, for the difference $(-4) - (+3)$:

$(\bullet\bullet\bullet\bullet) - (\bigcirc\bigcirc\bigcirc)$

The first term has four black (negative) counters, and the second term has three white (positive) counters. How can you subtract white counters from black counters? Using the zero principle, $(+1) + (-1) = 0$. So you can add as many zeros (one white counter + one black counter) to the first term as you need.

$(\bullet\bullet\bullet\bullet \ \bullet\oslash \ \bullet\oslash \ \bullet\oslash) - (\oslash\oslash\oslash)$

Subtract three white counters to obtain the answer: 7 black counters, or -7.

4. Use the key to crack the code. Write the letter of the correct answer under the question to discover the rule for subtracting integers. The first one is done for you.

Key: | $-8 = A$ | $-6 = D$ | $-4 = E$ | $-2 = H$ | $0 = I$ | $+2 = O$ | $+4 = P$ | $+6 = S$ | $+8 = T$ |

$$\begin{array}{cc} (-10) \\ \underline{-(-2)} \\ A \end{array} \quad \begin{array}{c} (-3) \\ \underline{-(+3)} \end{array} \quad \begin{array}{c} (+2) \\ \underline{-(+8)} \end{array} \quad\quad \begin{array}{c} (+4) \\ \underline{-(-4)} \end{array} \quad \begin{array}{c} (+6) \\ \underline{-(+8)} \end{array} \quad \begin{array}{c} (-7) \\ \underline{-(-3)} \end{array}$$

$$\begin{array}{c} (-6) \\ \underline{-(-8)} \end{array} \quad \begin{array}{c} (+9) \\ \underline{-(+5)} \end{array} \quad \begin{array}{c} (-6) \\ \underline{-(-10)} \end{array} \quad \begin{array}{c} (-9) \\ \underline{-(-11)} \end{array} \quad \begin{array}{c} (+12) \\ \underline{-(+6)} \end{array} \quad \begin{array}{c} (-4) \\ \underline{-(-4)} \end{array} \quad \begin{array}{c} (+10) \\ \underline{-(+2)} \end{array} \quad \begin{array}{c} (-2) \\ \underline{-(+2)} \end{array}$$

6.7 Using Number Lines to Subtract Integers

▶ **GOAL:** Calculate the difference between integers using a number line.

1. The start and end points of arrows on a number line are given below. Write the subtraction question each arrow represents. Calculate each difference. Part (a) is done for you.

At-Home Help

To subtract integers on a number line, do the following:
- Draw an arrow from the position of the second term to the position of the first term.
- If the arrow points right, the answer is positive. If the arrow points left, the answer is negative.
- To find the answer, count the number of units in the arrow and give it the appropriate sign.

For example, to calculate $(+4) - (-3)$, the arrow will start at (-3) and end at $(+4)$. The arrow will be 7 units long, and point to the right. So the answer is $(+7)$.

a) Start of arrow: -2 End of arrow: $+4$

   Subtraction question: _____$(+4) - (-2)$_____

   Difference: ___$(+6)$___

b) Start of arrow: $+7$ End of arrow: -11

   Subtraction question: _____

   Difference: _____

c) Start of arrow: -26 End of arrow: -14

   Subtraction question: _____

   Difference: _____

2. Find the difference between the integers on a number line.

a) $(-32) - (+4)$

   The difference is _____.

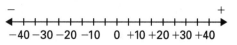

b) $(-32) - (-4)$

   The difference is _____ .

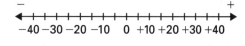

c) $(+32) - (+4)$

   The difference is _____ .

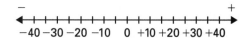

d) $(+32) - (-4)$

   The difference is _____ .

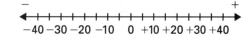

e) $(-16) - (-28)$

   The difference is _____ .

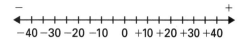

f) $(+14) - (+21)$

   The difference is _____ .

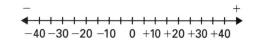

Solve Problems by Working Backward

▶ **GOAL:** Use the strategy of working backward to solve problems.

1. Work backward to find the original integer.

   a) • Subtract −6.
      • Find the opposite.
      • Add −2.
      • Subtract −4.
      • Find the opposite.
      • The answer is −5.
      • The original integer is _____ .

   b) • Add +14.
      • Add −5.
      • Find the opposite.
      • Subtract +6.
      • Add +3.
      • The answer is 0.
      • The original integer is _____ .

At–Home Help
When you are working backward, if the question asks you to add, do the opposite and subtract. If the question asks you to subtract, do the opposite and add.

2. An elevator goes down seven floors, up three floors, up six floors, down eight floors, up one floor, and down six floors. It is now on floor 3. Which floor did it start on?

3. Meagan bought a book for $12.00, received an allowance of $15.00, paid back a loan of $7.00, bought lunch for $8.50, earned $8.00 raking leaves, and found $2.00. She now has $22.25. How much did she start with?

4. Miguel and Yoshi went rock climbing. Miguel climbed up 15 m, back down 4 m, climbed 6 m, slid back 17 m, climbed up 22 m, rested and climbed 14 m more. Yoshi climbed 12 m, slid back 4 m, rested, climbed up 33 m and slid back 7 m. They are now together at the 40 m mark. At what level did each start climbing?

5. Shailini's plane leaves at 3:35 P.M. She has to check in 2 h before take-off. She takes 1 h and 35 min to get to the airport from her home. Shailini must first drop off her dog, Fido, at her friend's house next door. She wants to spend 45 min talking to her friend. What is the latest time she should leave her house?

Test Yourself

1. Which integer is greater?

   a) −4, −14   d) +7, +3

   b) +2, −2   e) −16, +2

   c) 0, +3   f) −8, +13

2. For each group of integers, draw a small number line in the space to the right. Mark the integers in order on the number line.

   a) −3, 0, −1, +5, −5

   b) +10, −5, −20, −15, +20

   c) −13, −7, +9, +4, −5

3. Order these integers from least to greatest.

   −5, +17, −4, 0, +2, −17, +8, +1

   ___, ___, ___, ___, ___, ___,

   ___, ___

4. Which makes each statement true: >, <, or = ?

   a) (−7) ___ (−6)

   b) (+3) ___ (−2)

   c) (+13) ___ (−5)

   d) 0 ___ (−5)

   e) (+7) ___ (+5) + (+2)

   f) (−6) + (+3) ___ (−3)

   g) (−5) − (−3) ___ (−4)

   h) (+8) + (−15) ___ (+8) − (−15)

   i) (−5) + (−6) ___ (−11) − (−6)

   j) (+4) − (+7) ___ (−1) + (−2)

5. Continue each pattern.

   a) +5, +2, −1, −4, ___, ___, ___

   b) −11, −7, −3, +1, ___, ___, ___

   c) +2, −3, +4, −5, ___, ___, ___

   d) 0, −2, +1, −1, +2, ___, ___, ___

6. Add these integers.

   a) (−8) + (+16) = _____

   b) (−9) + (−6) = _____

   c) (−3) + (+7) = _____

   d) (+5) + (−5) = _____

   e) (−23) + (+8) = _____

   f) (+90) + (−50) = _____

   g) (−17) + (−23) = _____

   h) (+40) + (−25) = _____

   i) (−2) + (+100) = _____

   j) (+33) + (−18) = _____

7. Subtract these integers.

   a) (−8) − (+16) = _____

   b) (−9) − (−6) = _____

   c) (−3) − (+7) = _____

   d) (+5) − (−5) = _____

   e) (−23) − (+8) = _____

   f) (+90) − (−50) = _____

   g) (−17) − (−23) = _____

   h) (+40) − (−25) = _____

   i) (−2) − (+100) = _____

   j) (+33) − (−18) = _____

8. Enter + or −.

   a) $(+5)$ ___ $(-6) = -1$

   b) $(+5)$ ___ $(+6) = -1$

   c) (-5) ___ $(-6) = +1$

   d) (-5) ___ $(+6) = +1$

9. Use the strategy of your choice to find the sums and differences below.

   a) $(-4) + (+10) + (+4) + (-2) =$ _____

   b) $(+5) + (-8) + (+3) + (-7) =$ _____

   c) $(-19) + (+14) + (+21) + (-23) =$ _____

   d) $(+5) - (-10) - (+4) =$ _____

   e) $(-3) - (-3) - (-3) =$ _____

   f) $(+26) - (-32) - (+15) - (-8) =$ _____

   For Questions 10, 11, and 12, write a number sentence using integers, and then solve.

10. The temperature at 5:00 A.M. was −14°C. By 5:00 P.M., it had risen to +8°C.

    What is the difference in the temperatures?

    _____ = _____°C

11. The temperature at 6:00 A.M. was −10°C. By noon, it had risen by 14°C.

    What was the temperature at noon?

    _____ = _____°C

12. The temperature at 2:00 P.M. was 19°C. By midnight, it was −3°C.

    What is the difference in the temperatures?

    _____ = _____°C

13. a) Draw counters in the space below to model and add $(-4) + (+2)$.

    b) Draw counters in the space below to model and subtract $(-6) - (+5)$.

14. Yuki made cookies on Monday. Some friends came over and ate half the cookies on Tuesday. On Wednesday, Yuki ate three cookies, and on Thursday, she took two cookies to school. By Friday, Yuki had seven cookies left. How many did she bake on Monday?

15. A student has $45 in her piggy bank. She receives $25 for her birthday and $15 for babysitting. She buys a pair of shoes for $40. How much money does she have left? Solve by first writing a number sentence using integers and then finding the best way to simplify addition.

    _____ = _____

7.1 Comparing Positions on a Grid

▶ **GOAL:** Locate positions on a Cartesian grid with integer coordinates.

1. Name the coordinates for each point. The first one is done for you.

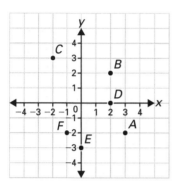

A: _____(3, −2)_____

B: _____

C: _____

D: _____

E: _____

F: _____

At–Home Help

Here are some tips to help you plot points on a grid.

- The horizontal number line on a grid is the *x*-axis, and the vertical number line is the *y*-axis.
- On the *x*-axis, the numbers to the right of zero are positive, and the numbers to the left of zero are negative. On the *y*-axis, the numbers above zero are positive, and the numbers below zero are negative.
- The first number in a set of coordinates refers to the position of a point relative to the *x*-axis. The second number refers to its position relative to the *y*-axis. For example, if a point has the coordinates (−3, 4), the point is located 3 units to the left of zero and 4 units above zero.

2. Name the coordinates for each point.

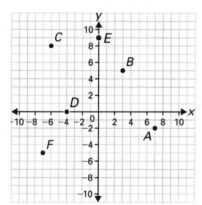

A: _____

B: _____

C: _____

D: _____

E: _____

F: _____

3. Plot the following points on the grid:

$A(−9, −9)$, $B(9, 9)$, $C(−9, 9)$, $D(9, −9)$, $E(0, −10)$, $F(4, −5)$, $G(2, −7)$, $H(−5, 5)$, $I(7, 6)$, $J(−8, −4)$

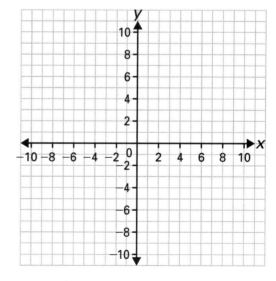

7.2 Translations

▶ **GOAL: Recognize the image of a 2-D shape after a translation.**

1. Write the new coordinates after each translation.

   a) $A(-1, 3)$ translated 1 unit to the left and 1 unit down _____

   b) $B(0, 0)$ translated 4 units to the right and 5 units up _____

   c) $C(-3, -7)$ translated 2 units to the right and 6 units up _____

2. A square has points $A(-2, -2)$, $B(-2, 2)$, $C(2, 2)$ and $D(2, -2)$. The square is translated 4 units to the right and 3 units down. What are the points of the image square?

   A': _____ C': _____ B': _____ D': _____

3. Translate triangle *DEF* 3 units to the right and 3 units up.

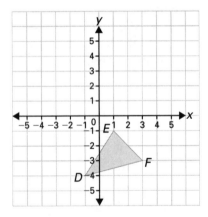

   Image coordinates:

   _____

   _____

   _____

4. Describe each transformation.

   a)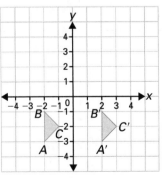

   b)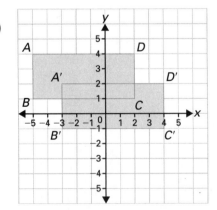

At–Home Help

A **translation** is the result of a slide along straight lines (left or right, up or down). For example, the triangle below on the left has points $A(-3, 1)$, $B(-2, 3)$, and $C(-1, 1)$. The diagram shows this triangle being translated 4 units to the right, or +4, and 1 unit down, or −1.

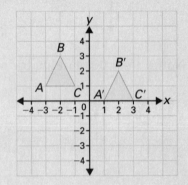

The new triangle is called the **image**. You can find the points of the image by adding +4 to each *x*-coordinate, and −1 to each *y*-coordinate. The new points are $A'(1, 0)$, $B'(2, 2)$, and $C'(3, 0)$.

7.3 Reflections

▶ **GOAL: Explore the properties of reflections of 2-D shapes.**

1. Reflect rectangle *ABCD* in the *y*-axis. Find the coordinates of the image.

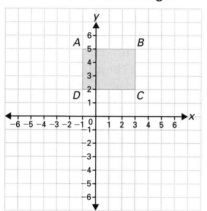

Image coordinates:

_____

_____

_____

_____

At–Home Help

A **reflection** is the result of a flip of a 2-D shape. Each point in the 2-D shape flips to the opposite side of the line of reflection, but stays the same distance from the line. The example below shows triangle *ABC* reflected in the *x*-axis to produce triangle *A'B'C'*.

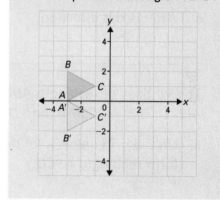

2. **a)** On the grid below, draw a rectangle with the coordinates *J*(3, 5), *K*(5, 5), *L*(5, −4) and *M*(3, −4).

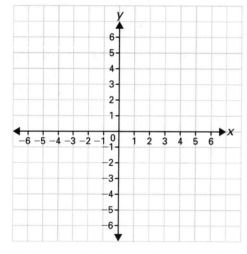

b) Draw the image of rectangle *JKLM* when reflected in the *x*-axis. Label the image *J'K'L'M'*. What are the coordinates of the image?

c) Draw the image of rectangle *JKLM* when reflected in the *y*-axis. Label the new image *J''K''L''M''*. What are the coordinates of this image?

3. Triangle *QRS* has the coordinates *Q*(2, −4), *R*(3, −1), and *S*(5, −3). What are the coordinates of the image if triangle *QRS* is reflected in the *x*-axis?

Q': _____   *R'*: _____   *S'*: _____

7.4 Rotations

▶ **GOAL: Identify the properties of a 2-D shape that stay the same after a rotation.**

1. Triangle *XYZ* has been rotated to give triangle *X'Y'Z'*.

a) What point is the centre of rotation?

_____

b) What is the angle of rotation?

_____

c) What is the direction of rotation?

_____

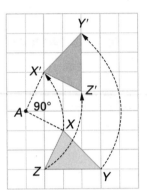

At-Home Help

The **centre of rotation** is a fixed point around which other points in a shape rotate in a clockwise (cw) or counterclockwise (ccw) direction. The centre of rotation may be inside or outside the shape.

To rotate a shape on a Cartesian grid, follow these steps for each point in the shape:

• Draw a line from the point to the centre of rotation.

• Use a protractor to measure the angle of rotation (for example, 90° in a clockwise direction). Draw another line segment at this angle.

• Place the point of a compass on the centre of rotation, and place the pencil on the point. Draw an arc from the point to meet the new line segment. Mark the new image point here.

• Repeat these steps for each point in the shape.

2. The vertices of rectangle *ABCD* have coordinates *A*(2, 4), *B*(3, 4), *C*(3, 1), and *D*(2, 1).

a) On the grid below, plot the points and join them to form rectangle *ABCD*.

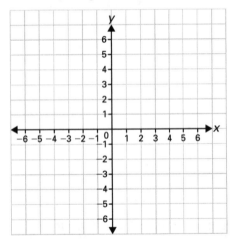

b) Determine the coordinates of rectangle *ABCD* after a 90° cw rotation about the origin, (0, 0).

c) Determine the coordinates of rectangle *ABCD* after a 90° ccw rotation about the origin.

d) Determine the coordinates of rectangle *ABCD* after a 180° cw rotation about the origin.

Congruence and Similarity

▶ **GOAL:** Investigate the conditions that make two shapes congruent or similar.

1. Are these two shapes congruent or similar? Explain your answer.

_____

_____

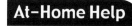

At-Home Help

Two figures are **similar** if they have exactly the same shape. They do not have to be the same size.

Two figures are **congruent** if they have exactly the same shape and size. Congruent figures can be placed one on top of the other so that they coincide.

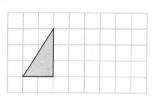

2. Look at the figures below.

   a) Find one pair of congruent figures. _____

   b) Find one pair of similar figures. _____

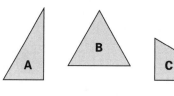

3. For each figure, draw a second figure beside it that is congruent.

   a) 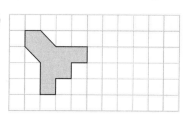   **b)**

4. For each figure, draw a second figure beside it that is similar.

   a) 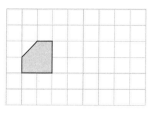   **b)**

5. a) The two triangles below are similar. Measure the corresponding sides and angles.

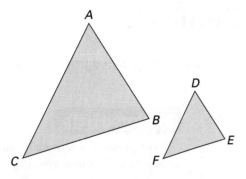

AB =	angle ABC =	DE =	angle DEF =
BC =	angle BAC =	EF =	angle EDF =
AC =	angle ACB =	DF =	angle DFE =

b) What do you notice about the sides and angles of the two triangles?

_____

_____

6. a) The two quadrilaterals below are congruent. Measure the corresponding sides and angles.

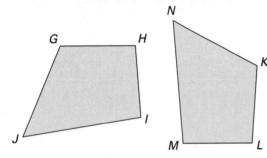

GH =	angle GHI =	KL =	angle KLM =
HI =	angle HIJ =	LM =	angle LMN =
IJ =	angle IJG =	MN =	angle MNK =
JG =	angle JGH =	NK =	angle NKL =

b) What do you notice about the sides and angles of the two quadrilaterals?

_____

_____

7.6 Tessellations

▶ **GOAL:** Create and analyze designs that tessellate a plane.

1. Look at the following diagram of a tessellation. This tessellation is made from one pentomino that has been transformed into different orientations.

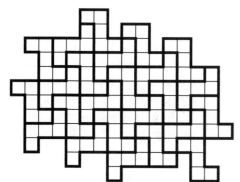

a) How many different orientations can you see in the tessellation? _____

b) Sketch the different orientations in the space below.

c) Choose one of the orientations from part (b). Describe how you would transform this orientation into a different orientation.

_____

2. Four square tiles can be used to make tetrominoes like these:

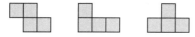

a) You can use tetrominoes to make tessellations. Choose one of the tetrominoes above. Use it to make a tessellation.

b) How many orientations are there in your tessellation? _____

c) What transformations did you use to produce the different orientations?

_____

Communicating about Geometric Patterns

▶ **GOAL:** Describe designs in terms of congruent, similar, and transformed images.

1. Describe each design. Assume that each square in the grid is 1 cm by 1 cm.

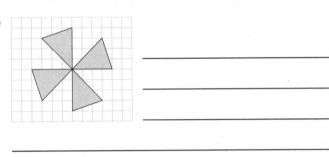

a)

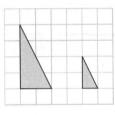

_____

_____

_____

_____

_____

b)

_____

_____

_____

_____

_____

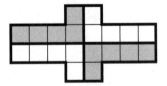

2. Describe each design, keeping the Communication Checklist in mind.

a)

b)

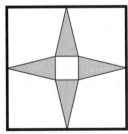

_____

_____

_____

_____

_____

▶ **GOAL:** Use transformations and properties of congruent shapes to solve problems.

1. Circle the letters of the shapes that are regular polygons.

A   B   C   D   E

2. You can determine a vertex angle of a polygon by tessellating the polygon around a single point. The diagram to the right shows a tessellation of four squares around a single point. What is the vertex angle of each square? (**Hint:** Remember that the total of all the angles around a single point is 360°.) _____

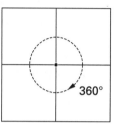

360°

3. **a)** The pentagon shown here is a regular polygon. Does this mean that you can tessellate it around a single point with no overlap or gaps? Explain your answer.

_____

_____

b) Test your answer by tracing the pentagon and trying to tessellate it around a point. What do you observe?

_____

c) The vertex angles of a regular pentagon are all 108°. How could you use this information to answer part (a)?

_____

_____

4. Trace triangle B from question 1. Tessellate it around the point below. What is the vertex angle of each triangle?

•

7.9 Tessellating Designs

▶ **GOAL:** Create irregular tiles, and determine whether irregular tiles can be used to tessellate a plane.

1. a) In the space below, create a tile by following this description:

- Start with a square.

- Change half of one side. Rotate this change 180° about the midpoint of the side to complete the side.

- Change the other sides the same way.

b) Use tracing paper to copy the tile and create a tessellation.

2. a) In the space below, create a tile by following this description:

- Start with an equilateral triangle.

- Change one side of the triangle. Rotate this change about each vertex to change the other two sides of the triangle in the same way.

b) Use tracing paper to copy the tile and create a tessellation.

Test Yourself

1. a) Name the coordinates for each point.

A: _____

B: _____

C: _____

D: _____

E: _____

F: _____

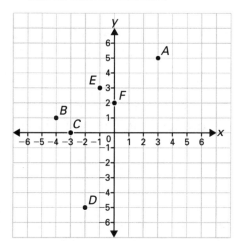

b) On the grid in part (a), plot these points: G(2, 2), H(−4, −1), I(5, −1) and J(0, 4).

2. Fill in the blanks by writing "above" or "below."

a) (3, −5) is _____ (5, −3).

b) (−2, 7) is _____ (4, 4).

3. Fill in the blanks by writing "left" or "right."

a) (5, −4) is to the _____ of (0, 3).

b) (−13, 3) is to the _____ of (−15, −4).

4. The vertices of rectangle *ABCD* have coordinates A(−2, 4), B(3, 4), C(3, −2) and D(−2, −2). Rectangle *ABCD* is translated 3 units to the left and 2 units up. Determine the coordinates of the image rectangle.

A': _____   C': _____

B': _____   D': _____

5. The vertices of triangle *EFG* have coordinates E(0, 3), F(3, 0), and G(5, 4).

a) Draw triangle *EFG* on the grid below.

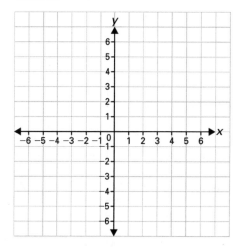

b) Triangle *EFG* is translated 5 units to the left and 3 units down. Draw the translation on the grid, and find the new coordinates.

E': _____

F': _____

G': _____

6. Reflect quadrilateral *HIJK* in the *x*-axis. Determine the image coordinates.

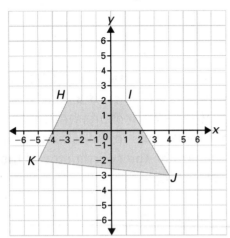

H': _____ *J*': _____

I': _____ *K*': _____

7. a) Rotate triangle *LMN* 180° cw about the origin, (0, 0). Label the image coordinates.

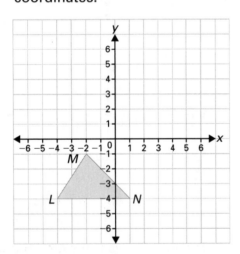

b) Predict a different rotation that would move triangle *LMN* to the same image as in part (a).

_____

8. Look at the figures below.

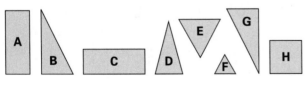

a) List two pairs of congruent shapes.

_____

b) List one pair of similar shapes.

_____

9. a) Use the pentomino to draw a tessellation on the grid below.

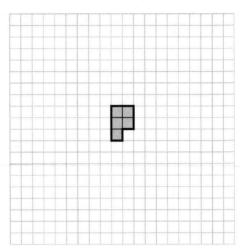

b) How many orientations did you use in the tessellation?

_____

c) List the transformations you used to produce the different orientations.

_____

_____

10. Use the Communication Checklist from Lesson 7.7 to write a description of the figure below.

_____

_____

_____

_____

_____

_____

_____

Exploring Pattern Representations

▶ **GOAL: Explore different ways of describing a pattern.**

Use the following pattern of squares to answer the questions below.

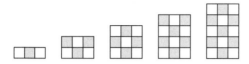

1. a) Count the number of squares in each figure. Fill out the table of values below for this sequence.

Term number	1	2	3	4	5
Term value					

b) Write a pattern rule for this sequence.

_____

2. a) Count the number of shaded squares in each figure. Fill out the table of values below for this sequence.

Term number	1	2	3	4	5
Term value					

b) Write a pattern rule for this sequence.

_____

3. a) Use the grid to create a scatter plot that represents both tables of values. Draw each line in a different colour.

b) Use your scatter plot to predict what the pattern of the 10th term will be.

How many squares in total? _____

How many shaded squares? _____

Using Variables to Write Pattern Rules

▶ **GOAL:** Use numbers and variables to represent mathematical relationships.

1. a) What stays the same and what changes in the pattern below?

_____

b) Describe the pattern rule in words.

_____

_____

> **At-Home Help**
>
> - A **variable** is a letter or symbol, such as *a*, *b*, or *x*, that represents a number.
> - An **algebraic expression** is a combination of one or more variables; it may include numbers and operation signs. For example, $2 \times d + 5$ is an algebraic expression that could represent two times the figure number plus five.

c) Write an algebraic expression that describes the pattern.

_____

2. a) Fill out the table of values for the pattern of circles.

Term number	1	2	3	4	5
Term value					

b) Describe how the number of circles (or term value) is related to the figure number (or term number).

_____

_____

c) Write an algebraic expression for the number of circles.

_____

3. Omar says this pattern can be described by the expression $2 + (n + 1)$. Tynessa says the pattern can be described by the expression $3 + n$.

a) Explain Omar's reasoning.

_____

_____

b) Explain Tynessa's reasoning.

_____

_____

4. a) Write an algebraic expression for the number of squares you would need to build the figures in this pattern.

_____

b) Write an algebraic expression for the number of triangles you would need to build the figures in this pattern.

_____

c) Write an algebraic expression for the number of circles you would need to build the figures in this pattern.

_____

5. Look at the following pattern.

a) Write an algebraic expression for the number of squares you would need to build the figures in this pattern. _____

b) Write an algebraic expression for the number of triangles you would need to build the figures in this pattern. _____

c) Add the two algebraic expressions together to form an expression for the total number of blocks you would need to build the figures in this pattern. _____

6. a) Fill out the table of values for the pattern below.

Figure number	1	2	3	4	5	...	10
Number of white squares							
Number of shaded squares							

b) Write an algebraic expression for the number of shaded squares in each figure. _____

c) Write an algebraic expression for the total number of squares in each figure. _____

8.3 Creating and Evaluating Expressions

▶ **GOAL: Translate statements into algebraic expressions, and evaluate the expressions.**

1. Each expression represents any number in a sequence. Replace n with 1, 2, 3, and 4 to calculate the first four numbers in each sequence.

   a) $n + 8$ _____, _____, _____, _____

   b) $8n$ _____, _____, _____, _____

   c) $8 - n$ _____, _____, _____, _____

   d) $\dfrac{12}{n}$ _____, _____, _____, _____

   e) $2n + 3$ _____, _____, _____, _____

2. Evaluate each algebraic expression when $a = 2$ and $b = 6$.

   a) $5a$

   b) $3b$

   c) $2 + a$

   d) $12 - b$

   e) $2a + 3$

   f) $4(b - 3)$

   g) $10 - 5a$

   h) $(a + 1)^2$

3. Rana uses the formula $1.75p$ to calculate, in dollars, the cost of p packages of pens.

   a) What is the cost of 10 packages of pens? _____

   b) What is the cost of 1 package of pens? _____

4. Colin works in a china shop. He calculates the price, in dollars, of a gift-wrapped purchase using the formula $c + 2$. In his formula, c is the cost of the china.

   a) What is the total price of a $19 gift-wrapped item? _____

   b) What is the total price of a $100 gift-wrapped item? _____

5. Kaitlyn buys books from a book club. She calculates the total cost, in dollars, of each order using the formula $15b + 5$. In her formula, b represents the number of books she orders.

   a) What is the total cost of ordering four books? _____

   b) What is the total cost of ordering two books? _____

   c) What is the total cost of ordering one book? _____

6. Write an algebraic expression for each cost.

   a) hot chocolate at $2 per cup

   b) $10 per pizza plus $2 delivery cost

   c) jeans at $35 each minus a one-time $10 discount

7. a) Write an algebraic expression to represent the cost of buying shoes online, at $20 each pair, plus $5 shipping for the whole order.

   b) What is the total cost of buying 3 pairs of shoes online? _____

   c) What is the total cost of buying 10 pairs of shoes online? _____

8. Evaluate $3(x + 4)$ when $x = 5$. Show your steps.

9. Yuki runs 0.5 km to the park every morning. She runs around the 2 km park trail p times, then runs 0.5 km back home.

   a) Write an algebraic expression for the distance Yuki runs.

   b) How far does Yuki run if she goes around the park once? _____

   c) How far does Yuki run if she goes around the park four times? _____

8.4 Solving Equations by Inspection

▶ **GOAL: Write equations and solve them by inspection.**

1. Solve each equation by inspection.

   a) $x + 1 = 6$

   $x =$ _____

   b) $3a = 9$

   $a =$ _____

   c) $8 - p = 6$

   $p =$ _____

   d) $5 + b = 16$

   $b =$ _____

   e) $2z - 1 = 3$

   $z =$ _____

   f) $4 + 3d = 19$

   $d =$ _____

2. Ravi solved an equation by the following steps:

   $4t - 8 = 16$

   $4t = 8$

   $t = 2$

   a) Check Ravi's solution.

   Is Ravi's solution correct? _____

   b) If your answer to part (a) is no, solve the equation by inspection.

3. **a)** Write an equation that describes the number of triangles needed to build each figure in the pattern below. _____

   b) Suppose that you want to build a figure in this pattern using 15 triangles. Write the equation you must solve to determine the figure number. _____

   c) Solve your equation.

   d) Use your equation to determine which figure number you could build using 21 triangles.

Solving Equations by Systematic Trial

▶ **GOAL: Write equations and solve them using systematic guessing and testing.**

1. Fill in each table to solve the equation. In the second column, show your work. In the third column, write "too high," "too low," or "correct."

   a) $y + 5 = 12$

Predict y.	Evaluate $y + 5$.	Is this the correct solution?
5		
10		
7		

   b) $3m = 333$

Predict m.	Evaluate $3m$.	Is this the correct solution?
200		
150		
111		

   c) $5r - 10 = 95$

Predict r.	Evaluate $5r - 10$.	Is this the correct solution?
15		
21		
25		

At-Home Help

To solve an equation using systematic trial, follow these steps:
- Start by guessing what the variable might be. For example, to solve the equation $3x + 2 = 47$, you could guess that $x = 20$. Check your guess: $3(20) + 2 = 62$ (too high).
- Guess again. Your second guess might be $x = 10$. Check your guess. For this example, $3(10) + 2 = 32$ (too low).
- Keep guessing until you find the answer. Your third guess might be $x = 15$. Check your guess: $3(15) + 2 = 47$ (correct!). The solution is 15.

2. Use systematic trial to find the value of each variable.

   a) $11 + x = 75$

   b) $q - 14 = 102$

   c) $9w = 153$

   d) $5 + 7c = 89$

   e) $26 = 5 + 3e$

   f) $8k - 16 = 200$

   g) $437 + s = 488$

   h) $525 - 6u = 339$

3. Write each sentence as an equation. Solve each equation.

   a) Four times a number plus 100 is 140.

   b) Seven times a number is 294.

   c) Four times a number minus 52 is 212.

4. Tynessa solved an equation by systematic trial in the following way:

   $24 + 12z = 180$

Predict z.	Evaluate 24 + 12z.	Is this the correct solution?
20	$(24 + 12) \times 20 = 720$	too high
2	$(24 + 12) \times 2 = 72$	too low
5	$(24 + 12) \times 5 = 180$	correct

   a) What mistake did Tynessa make? _____

   _____

   b) Fill out the table below to solve the equation.

Predict z.	Evaluate 24 + 12z.	Is this the correct solution?

5. The formula for the area of a triangle is $A = \dfrac{hb}{2}$.

   a) A triangle has $h = 3$ units and $b = 4$ units. What is
      the area, A?

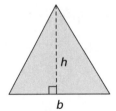

   b) A triangle has $A = 24$ units squared and $b = 12$ units. What is the
      height, h?

   c) A triangle has $A = 92$ units squared and $h = 23$ units. What is the
      base, b?

Communicating the Solution for an Equation

▶ **GOAL: Communicate about solving equations using correct mathematical language.**

1. Explain what is happening in each step of the following solution.

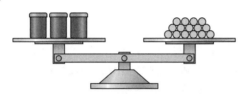

$$3c = 15$$
$$3c \div 3 = 15 \div 3$$
$$c = 5$$

Explanation: _____

_____

_____

At-Home Help

Use this Communication Checklist to help you communicate about solving equations.

Communication Checklist
☑ Show each step of your thinking.
☑ Express yourself clearly.
☑ Use appropriate mathematical language, like *variable*, *solution*, and *algebraic equation*.

2. For each balance problem, use pictures and words to explain each step in the solution.

a)

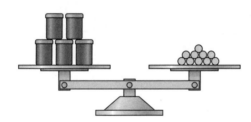

b)

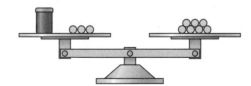

c)

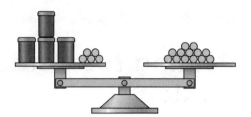

3. Improve Omar's explanation:

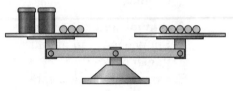

"I divided everything by 2 because there are two containers. First, I subtracted 3 from each side to make them equal."

Your explanation:

$2m + 3 = 5$
$m = (5 - 3) \div 2$
$ = 1$

4. Find and explain any errors in Tynessa's solution to this balance problem. Rewrite the solution correctly, and check your work.

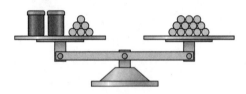

$2c + 6 = 12$
$c + 6 = 12 \div 2$
$c + 6 = 6$
$ c = 6 + 6$
$ c = 12$

Explain any errors you found. _____

_____

Your solution:

Test Yourself

1. **a)** Use words to describe the pattern rule for the total number of blocks in each figure below.

   _____

   _____

   b) Write an algebraic expression to describe the pattern rule.

   _____

2. Use each description to create an algebraic expression for the pattern rule.

   a) The number of blocks triples each time.

   Algebraic expression:

   _____

   b) The first figure has four blocks. The number of blocks increases by one each time.

   Algebraic expression:

   _____

   c) The first figure has two triangles and one circle. The number of triangles increases by one each time. The number of circles stays the same.

   Algebraic expression:

   _____

3. Each algebraic expression represents a pattern rule. Draw examples that show possible figures for the first three terms of each pattern.

   a) $x + 5$

   b) $2z$

   c) $4p + 1$

   d) $3v + 2$

4. Evaluate each expression for the given variable.

   a) $a + 4$, where $a = 5$

   b) $7b$, where $b = 2$

   c) $10 - c$, where $c = 8$

   d) $9d + 1$, where $d = 1$

5. Paul works at a hamburger stand. He earns $15 per day, plus $1 for every hamburger he sells.

   a) Write an algebraic expression that describes how much money Paul makes.

   _____

   b) How much money will Paul make if he sells 25 hamburgers in one day?

   c) How much money will Paul make if he sells 100 hamburgers in one day?

6. Solve each equation by inspection.

   a) $x + 7 = 19$ **c)** $2m + 1 = 5$

   b) $5p = 45$ **d)** $3b - 1 = 17$

7. **a)** Write an equation that describes the number of triangles needed to build each figure in this pattern.

   _____

   b) Suppose you want to build a figure in this pattern using 16 triangles. Write the equation you must solve to determine the figure number.

   _____

   c) Solve your equation.

   d) Check your equation.

8. Fill out the table to find the value of the variable in the equation $4 + 2k = 106$.

Predict k.	Evaluate $4 + 2k$.	Is this correct?

9. **a)** Write an equation for this balance problem.

   _____

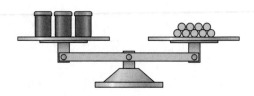

   b) Solve your equation.

   c) Explain your solution.

   _____

   _____

10. Write a clear algebraic solution for each balance problem. Check your work.

   a) $x + 4 = 9$

   b) $2x = 6$

   c) $3x + 1 = 13$

   d) $4x - 11 = 5$

Adding Fractions with Pattern Blocks

▶ **GOAL: Add fractions that are less than 1 using concrete materials.**

1. Shade $\frac{1}{4}$ of each diagram.

2. Draw two different diagrams to show $\frac{2}{5}$.

3. **a)** Shade $\frac{1}{6}$ of the diagram.

   b) In the same diagram,

   shade another $\frac{1}{6}$.

   c) $\frac{1}{6} + \frac{1}{6} = \dfrac{}{6} = \dfrac{}{3}$

4. **a)** Shade $\frac{1}{2}$ of the diagram.

   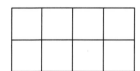

   b) Shade $\frac{1}{8}$ more of the diagram.

   c) $\frac{1}{2} + \frac{1}{8} = \dfrac{}{8}$

9.2 Adding Fractions with Models

▶ **GOAL: Add fractions that are less than 1 using fraction strips and number lines.**

1. Chang used fraction strips to calculate $\frac{2}{3} + \frac{1}{6}$, but he did not get the correct solution. Find and explain Chang's error. Then use fraction strips to rewrite the solution correctly.

   Chang's solution: "The number 6 is a common denominator of $\frac{2}{3}$ and $\frac{1}{6}$, because 6 is a common multiple of both 3 and 6. I divided each fraction strip into six parts. I coloured two rectangles on one of the strips, because 2 is the numerator of $\frac{2}{3}$. I coloured one rectangle on the second strip, because 1 is the numerator of $\frac{1}{6}$. I added the number of rectangles to get $\frac{3}{6}$, which is equal to $\frac{1}{2}$."

At-Home Help

These models can help you add fractions.
• A fraction strip is a strip of paper divided into rectangles that are the same size. The whole length of each fraction strip should be the same, no matter what fraction the strip represents.

1/4			

• A number line is like a thin fraction strip.

$$0 \quad \frac{1}{4} \quad \frac{1}{2} \quad \frac{3}{4} \quad 1$$

   Chang's error: _____

   _____

   Your solution: _____

2. Use fraction strips to add. Show your work.

   a) $\frac{1}{4} + \frac{1}{4} = \frac{2}{8}$

   b) $\frac{2}{3} + \frac{1}{3} = \frac{3}{3}$ or 1

   c) $\frac{3}{6} + \frac{1^{\times 3}}{2^{\times 3}} = \frac{3}{6} = \frac{3}{6} =$

   d) $\frac{1^{\times 3}}{5^{\times 3}} + \frac{2^{\times 5}}{3^{\times 5}} = \frac{3}{15} + \frac{5}{15} = \frac{13}{15}$

   e) $\frac{3}{8} + \frac{2^{\times 2}}{4^{\times 2}} = \frac{3}{8} + \frac{4}{8} = \frac{7}{8}$

   f) $\frac{1}{2} + \frac{1}{5}$

3. Sandra is trying to solve the sum $\frac{1}{5} + \frac{1}{4}$ using a number line. Read her explanation below, and then complete her solution using the number line. Explain your work.

Sandra's part of the solution: "I started by finding a common denominator for the two fractions: 20. Then I found equivalent fractions: $\frac{1}{5}$ becomes $\frac{4}{20}$, and $\frac{1}{4}$ becomes $\frac{5}{20}$."

$$\frac{1}{5} = \frac{(1 \times 4)}{(4 \times 4)} \qquad\qquad \frac{1}{4} = \frac{(1 \times 5)}{(4 \times 5)}$$
$$= \frac{4}{20} \qquad\qquad\qquad = \frac{5}{20}$$

"I drew the number line and divided it into 20 parts. I drew an arrow to show $\frac{4}{20}$."

0      $\frac{4}{20}$ 1

Your part of the solution: _____

_____

4. Use a number line to add. Show your work.

a) $\frac{1}{5} + \frac{1}{2}$ d) $\frac{1}{3} + \frac{1}{8}$

b) $\frac{5}{8} + \frac{1}{4}$ e) $\frac{2}{5} + \frac{2}{8}$

c) $\frac{3}{4} + \frac{2}{8}$ f) $\frac{1}{6} + \frac{2}{5}$

5. One hour before the game started, $\frac{1}{8}$ of the stadium was filled. In the next 30 min, another $\frac{1}{4}$ of the stadium was filled. What fraction of the stadium was filled 30 min before game time?

6. Ravi practised piano 35 min on Saturday, and $\frac{1}{6}$ of an hour on Sunday. What fraction of an hour did he practise in total?

9.3 Multiplying a Whole Number by a Fraction

▶ **GOAL: Use repeated addition to multiply fractions by whole numbers.**

1. Chang is using grids and counters to multiply $4 \times \frac{2}{3}$. Part of his solution is shown below.

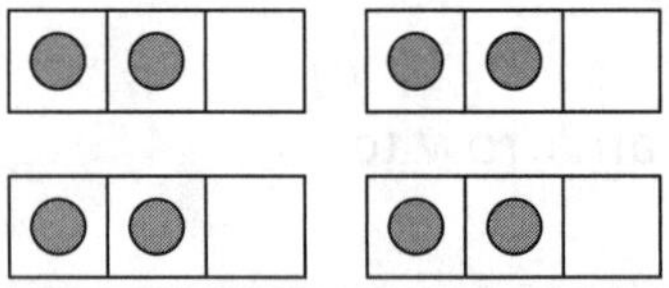

Chang's solution: "$4 \times \frac{2}{3}$ is four sets of $\frac{2}{3}$. I used 1-by-3 rectangles, since I want to show thirds, and $1 \times 3 = 3$. I showed four sets of $\frac{2}{3}$ by putting counters on two out of three squares in each of the four rectangles."

Complete Chang's solution.

a) How many squares are covered with a counter? _____

b) $4 \times \frac{2}{3}$ = _____

c) Write your answer from part (b) as a mixed number. _____

2. Multiply.

a) $3 \times \frac{1}{4}$

b) $2 \times \frac{3}{5}$

c) $5 \times \frac{1}{2}$

d) $7 \times \frac{1}{3}$

e) $4 \times \frac{5}{6}$

f) $10 \times \frac{2}{7}$

3. Replace each missing value with a single-digit number to make the sentence true.

a) $3 \times \dfrac{1}{\boxed{}} = \dfrac{3}{8}$

b) $\boxed{} \times \dfrac{2}{3} = \dfrac{4}{3}$, or $1\dfrac{1}{\boxed{}}$

c) $4 \times \dfrac{\boxed{}}{6} = \dfrac{12}{6}$, or $\boxed{}$

d) $\boxed{} \times \dfrac{\boxed{}}{5} = \dfrac{12}{5}$, or $\boxed{}\dfrac{2}{5}$

Subtracting Fractions with Models

▶ **GOAL: Subtract fractions less than 1 using fraction strips and number lines.**

1. Ryan used a number line to find the difference of $\frac{2}{3} - \frac{1}{4}$. Part of his solution is shown below.

   Ryan's solution: "I used a number line showing twelfths, since 12 is a common denominator for $\frac{2}{3}$ and $\frac{1}{4}$. I converted each fraction."

   $$\frac{2}{3} = \frac{(2 \times 4)}{(3 \times 4)} = \frac{8}{12}$$

   $$\frac{1}{4} = \frac{(1 \times 3)}{(4 \times 3)} = \frac{3}{12}$$

   "Then I drew an arrow to represent the fraction $\frac{2}{3}$, or $\frac{8}{12}$."

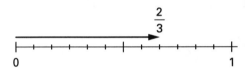

   Complete Ryan's solution using the number line. Explain your work.

2. Calculate each difference. Use a number line or fraction strips, and show your work.

   a) $\frac{3}{4} - \frac{1}{2}$

   b) $\frac{9}{10} - \frac{4}{5}$

   c) $\frac{4}{7} - \frac{1}{3}$

   d) $\frac{2}{5} - \frac{1}{6}$

   e) $\frac{9}{5} - \frac{2}{3}$

   f) $\frac{11}{4} - \frac{6}{7}$

3. Kwami started out with $\frac{3}{5}$ of a box of crackers. He ate $\frac{1}{3}$ more. What fraction of the box remains?

9.5 Subtracting Fractions with Grids

▶ **GOAL: Subtract fractions using grids and counters.**

1. Tynessa is using a grid and counters to calculate $\frac{3}{5} - \frac{1}{4}$. Part of her solution is shown below.

Tynessa's solution: "I used a 5-by-4 rectangle, since I wanted to show fifths and quarters. The rectangle has 20 squares since 20 is a common denominator of $\frac{3}{5}$ and $\frac{1}{4}$. Each row shows $\frac{1}{5}$. To show $\frac{3}{5}$, I covered three rows with counters."

"Each column shows $\frac{1}{4}$, so I moved the counters to fill as many columns as possible."

"I removed one complete column to model subtracting $\frac{1}{4}$."

Complete Tynessa's solution.

a) What is the final answer?

_____

b) Explain how you know:

_____

_____

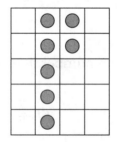

> **At–Home Help**
>
> You can use grids and counters to represent fractions, as you learned in Lesson 9.3. To subtract fractions, remember to use a grid that has the same number of squares as the common denominator. For example, if you want to calculate $\frac{2}{5} - \frac{1}{6}$, use a grid that is 5 by 6, or 30 squares.

2. Jody used a grid and counters to calculate $\frac{3}{7} - \frac{1}{3}$, but she did not get the correct answer. Find and explain Jody's error. Then rewrite the solution correctly.

Jody's solution: "I used a 7-by-3 rectangle, since I wanted to show sevenths and thirds. The rectangle has 21 squares, since 21 is the common denominator of 7 and 3. Each row shows $\frac{1}{7}$. To show $\frac{3}{7}$, I covered 3 rows with counters. This represents $\frac{9}{21}$."

Copyright © 2006 by Nelson Education Ltd.

"Next, I wanted to subtract $\frac{1}{3}$. Each column shows $\frac{1}{3}$, so I removed the counters from one column. Six of the squares still have counters, so the difference is $\frac{6}{21}$."

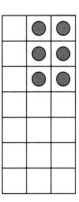

Jody's error: _____

_____

Your solution:

3. Use a grid and counters to model and calculate each difference.

a) $\frac{7}{6} - \frac{1}{4}$ e) $\frac{13}{12} - \frac{3}{4}$

b) $\frac{8}{3} - \frac{2}{5}$ f) $\frac{5}{4} - \frac{5}{8}$

c) $\frac{7}{8} - \frac{1}{2}$ g) $\frac{1}{5} - \frac{1}{7}$

d) $\frac{11}{9} - \frac{2}{3}$ h) $\frac{15}{8} - \frac{2}{5}$

4. Yuki made brownies, and ate $\frac{1}{5}$ of them. James ate another $\frac{2}{7}$ of the brownies.

a) What fraction of the brownies were eaten?

b) What fraction of the brownies were left?

5. Over one year, a music store sold $\frac{7}{8}$ of the total number of CDs in stock. During the year, $\frac{1}{6}$ of the total number of CDs in stock were returned. What fraction of the total number of CDs did the store make a profit from?

Adding and Subtracting Mixed Numbers

▶ **GOAL: Add and subtract mixed numbers using different models.**

1. Use a model to add.

   a) $1\frac{1}{3} + 2\frac{2}{3}$

   b) $2\frac{3}{4} + 5\frac{1}{2}$

   c) $4\frac{2}{3} + 3\frac{1}{6}$

   d) $1\frac{5}{8} + 7\frac{1}{4}$

   e) $5\frac{1}{6} + 10\frac{11}{12}$

   f) $3\frac{4}{5} + 1\frac{5}{6}$

   g) $8\frac{7}{9} + 4\frac{1}{2}$

   h) $6\frac{2}{7} + 2\frac{1}{3}$

> ### At-Home Help
>
> When adding and subtracting mixed numbers, you can use these models to help you:
> • grid and counters
> • fraction strips
> • number line
>
> When adding mixed numbers, you can add the whole numbers and the fractions separately, like this:
>
> $1\frac{1}{3} + 2\frac{1}{4} = (1 + 2) + (\frac{1}{3} + \frac{1}{4})$
>
> When subtracting a mixed number from a whole number, ask yourself, "How much do I need to add to the mixed number to get to the whole number?" For example, to calculate $4 - 2\frac{1}{3}$, adding $\frac{2}{3}$ and 1 to the mixed number will give you 4. So the solution is $1\frac{2}{3}$.

2. Use a number line or fraction strips to subtract.

   a) $5 - 2\frac{1}{4}$

   b) $3 - 1\frac{1}{5}$

   c) $6 - 2\frac{3}{7}$

   d) $4 - 3\frac{5}{8}$

   e) $7 - 5\frac{1}{2}$

   f) $11 - 9\frac{4}{9}$

   g) $8 - 1\frac{6}{7}$

   h) $2 - 1\frac{11}{12}$

3. Express $\frac{14}{12}$ and $\frac{31}{10}$ as mixed numbers. Then add the mixed numbers.

4. What was the total time spent on each exercise? Express your answers as mixed numbers.

		Day 1	Day 2	Total time
a)	swimming	$3\frac{1}{2}$ h	$1\frac{3}{4}$ h	
b)	running	$1\frac{1}{5}$ h	$2\frac{5}{6}$ h	
c)	kayaking	$4\frac{1}{4}$ h	$3\frac{7}{8}$ h	

5. James shopped for $2\frac{1}{3}$ h. He spent $\frac{1}{4}$ h travelling to the store, and $\frac{1}{4}$ h travelling back. How long was he gone?

6. a) Tynessa is 13 years old. How old was she $5\frac{1}{2}$ years ago?

   b) Tynessa's older brother is $18\frac{7}{8}$ years old. How old will he be in $4\frac{1}{4}$ years?

   c) Tynessa's little sister is 4 years old. How old was she $2\frac{3}{5}$ years ago?

7. Kwami and his friends ordered 7 pizzas. They ate $5\frac{11}{12}$ pizzas. How much pizza was left over?

8. Bonnie was away from home for 5 hours on Tuesday. She spent $1\frac{4}{9}$ h travelling, and the rest of the time studying at the library. How much time did she have to study?

9. Sandra bought 6 granola bars. She ate $1\frac{4}{5}$ on Tuesday, and $3\frac{1}{8}$ on Wednesday. How many of the granola bars does she have left?

9.7 Communicating about Estimation Strategies

▶ **GOAL: Explain how to estimate sums and differences of fractions and mixed numbers.**

1. Ryan bought 6 cases of juice for his party. People drank $4\frac{9}{12}$ cases of juice at the party. Ryan wants to estimate how much juice he has left.

   Ryan's solution: "I bought 6 cases of juice, and more than 4 cases were used up. So I think I have about 2 cases left."

   a) What is wrong with Ryan's estimation?

   _____

   _____

   b) Make your own estimation of how much juice Ryan has left. Use words and pictures to explain your work.

At-Home Help

Use this Communication Checklist to help you explain your work.

Communication Checklist
- ☑ Did you show all the necessary steps?
- ☑ Were your steps clear?
- ☑ Did you include words to describe your model, as well as pictures?
- ☑ Did your words support your use of the model?

2. Rana is making toffee. The recipe asks for $2\frac{3}{4}$ c. of white sugar, and $1\frac{1}{8}$ c. of brown sugar. About how much sugar is used in total? Use words and pictures to explain how to estimate the answer.

3. Miguel wants to cover the walls of his room with wood panelling. The wood panelling comes in large pieces. Each wall needs a different number of pieces, shown in the table below. About how many pieces of panelling does Miguel need in total?

North wall	$2\frac{1}{3}$ pieces
West wall	$\frac{1}{2}$ piece
South wall	$1\frac{4}{5}$ pieces
East wall	3 pieces

Adding and Subtracting Using Equivalent Fractions

▶ **GOAL: Develop a method for adding and subtracting fractions without using models.**

1. Find a common denominator for each pair of fractions. Then find equivalent fractions using that common denominator.

   a) $\frac{5}{8}$ and $\frac{3}{4}$

   c) $\frac{11}{12}$ and $\frac{1}{4}$

   b) $\frac{1}{2}$ and $\frac{2}{5}$

   d) $\frac{4}{7}$ and $\frac{4}{5}$

2. Fill in the missing values.

   a) $\frac{1}{2} - \frac{1}{4} = \frac{\boxed{}}{4} - \frac{1}{4}$

   $= \frac{\boxed{}}{4}$

   b) $\frac{2}{5} + \frac{2}{3} = \frac{\boxed{}}{15} + \frac{\boxed{}}{15}$

   $= \frac{\boxed{}}{15}$

   $= 1\frac{\boxed{}}{15}$

3. Add.

   a) $\frac{5}{7} + \frac{1}{14}$

   e) $\frac{7}{8} + \frac{3}{4}$

   b) $\frac{3}{8} + \frac{1}{4}$

   f) $\frac{3}{10} + \frac{5}{8}$

   c) $\frac{8}{9} + \frac{2}{3}$

   g) $\frac{1}{5} + \frac{1}{4}$

   d) $\frac{4}{7} + \frac{1}{6}$

   h) $\frac{1}{8} + \frac{1}{12}$

4. Subtract.

a) $\frac{5}{6} - \frac{2}{3}$

e) $\frac{1}{2} - \frac{1}{7}$

b) $\frac{3}{5} - \frac{1}{4}$

f) $\frac{3}{4} - \frac{1}{3}$

c) $\frac{9}{10} - \frac{2}{5}$

g) $\frac{6}{7} - \frac{4}{5}$

d) $\frac{1}{8} - \frac{1}{9}$

h) $\frac{9}{12} - \frac{3}{10}$

5. Romona painted $\frac{5}{6}$ h on Tuesday and $\frac{3}{4}$ h on Wednesday. How long did she paint in total? Write your answer as a mixed number.

6. A journey takes 3 days. You have travelled for $\frac{2}{3}$ of a day. How many days are left?

7. After a tough tennis game, Simon drank $\frac{3}{5}$ of a bottle of lemonade, and Indira drank $\frac{2}{3}$ of a bottle. Who drank more lemonade? How much more?

8. James studied for 2 h. He spent $\frac{6}{7}$ h studying history. How much time did he spend studying other topics?

9. Jody has finished $\frac{4}{5}$ of her homework. Sandra has finished $\frac{4}{7}$. How much more has Jody completed than Sandra?

10. Colin won $\frac{7}{10}$ of the board games he played with Kaitlyn.

a) What fraction of the games did Kaitlyn win? _____

b) How much more did Colin win than Kaitlyn? Explain below.

Test Yourself

1. Each diagram helps you add the fractions. Match the addition question to the diagram.

   a) $\frac{1}{5} + \frac{3}{5}$ A

   b) $\frac{1}{8} + \frac{6}{8}$ B

   c) $\frac{3}{10} + \frac{1}{10}$ C

2. Write each of these fractions in lowest terms.

   a) $\frac{6}{8}$ = _____

   b) $\frac{9}{12}$ = _____

   c) $\frac{4}{10}$ = _____

   d) $1\frac{2}{4}$ = _____

   e) $3\frac{8}{20}$ = _____

   f) $4\frac{2}{10}$ = _____

3. Fill in the fraction strips to add or subtract.

   a) $\frac{5}{6} + \frac{1}{3}$

   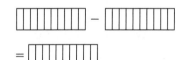

   b) $\frac{3}{5} - \frac{1}{2}$

4. Use the number lines to add or subtract.

   a) $\frac{2}{3} - \frac{5}{9}$

   b) $\frac{1}{3} + \frac{3}{8}$

5. Add or subtract.

   a) $\frac{4}{7} + \frac{3}{7}$

   b) $\frac{5}{8} - \frac{1}{8}$

   c) $\frac{9}{10} - \frac{4}{5}$

   d) $\frac{1}{6} + \frac{1}{8}$

   e) $\frac{2}{3} + \frac{1}{9}$

   f) $\frac{1}{2} - \frac{2}{7}$

   g) $\frac{7}{8} + \frac{1}{5}$

   h) $\frac{5}{6} - \frac{1}{10}$

6. Draw grids and counters to help you multiply. Show your work. Express your answers as mixed numbers in lowest terms.

   a) $5 \times \frac{2}{3}$

   b) $3 \times \frac{7}{10}$

7. Subtract $\frac{1}{4} - \frac{1}{12}$ using a grid and counters. Show and explain your work.

b) How old will she be in $4\frac{5}{12}$ years?

c) How old was she $3\frac{1}{5}$ years ago?

8. Add.

a) $3\frac{1}{2} + 1\frac{1}{3}$

c) $7\frac{1}{8} + 2\frac{3}{7}$

b) $6\frac{2}{3} + 5\frac{1}{4}$

13. Kaitlyn has five boxes of crackers. Each box is $\frac{1}{4}$ full.

a) If she combines all the crackers, how many full boxes of crackers will she have?

b) What fraction of a cracker box will she have left over? _____

9. Subtract.

a) $3 - 1\frac{2}{5}$

c) $9 - 4\frac{9}{10}$

b) $5 - 3\frac{1}{6}$

14. Indira cleaned $\frac{1}{6}$ of her room on Tuesday and $\frac{1}{4}$ of her room on Wednesday. What fraction of the room does she have left to clean?

15. Kaitlyn uses tubes of paint in art class. She started with 9 tubes of paint. First she painted a small picture that used $\frac{7}{8}$ of a tube of paint. Then she painted a larger picture that used up $3\frac{1}{6}$ tubes of paint.

10. Subtract using equivalent fractions.

a) $\frac{5}{8} - \frac{1}{4}$

c) $\frac{1}{6} - \frac{1}{9}$

b) $\frac{7}{9} - \frac{2}{3}$

a) About how much paint did Kaitlyn use in total?

11. Yuki was paid for five days of work. She spent $\frac{1}{3}$ of her pay. How much of her pay does she have left?

b) About how much paint does Kaitlyn have left?

12. Paul's grandmother is $75\frac{1}{2}$ years old.

a) How old will she be in $\frac{1}{5}$ of a year?

16. On Monday, it rained for $\frac{1}{3}$ of the day, snowed for $\frac{1}{5}$ of the day, and hailed for $\frac{1}{6}$ of the day. What fraction of the day had some form of precipitation?

Building and Packing Prisms

▶ **GOAL: Build prisms from nets.**

1. You can use plastic straws, pipe cleaners, and paper fasteners to construct skeletons of 3-D shapes.
   - Cut the pipe cleaners in half. Bend each piece to form a loop with the two ends together.
   - Push the two ends of a loop into one end of a straw. Push another loop into the other end of the straw. Continue until you have nine straws ready.

> **At-Home Help**
>
> A **prism** is a 3-D shape with opposite congruent bases; the other faces are parallelograms. The following materials can help you build models of prisms:
> - scissors, tape, and sheets of card, for building prisms from nets
> - straws and pipe cleaners, for constructing skeletons of prisms

   a) Use the paper fasteners to join three straws to make a triangle.

   How many sides does the triangle have? _____

   How many vertices does the triangle have? _____

   b) Make a second triangle. Join the two triangles together using three more straws to form the skeleton of a triangular prism. Sketch the prism in the space to the right.

   How many edges does the prism have? _____

   How many faces does it have? _____

   How many vertices does it have? _____

2. Copy each net onto a sheet of thin card. (You will need to make the nets bigger.) Cut out and construct the nets. For each prism, count the number of edges, faces, and vertices.

   a) rectangular prism

   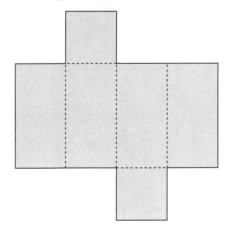

   b) pentagonal prism

   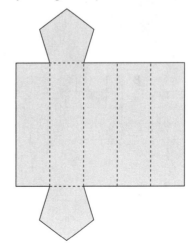

10.2 Building Objects from Nets

▶ **GOAL: Build 3-D shapes from nets.**

1. Circle the letter of each net that can be folded to make a cube.

   a)

   b)

   c)

   d)

   e)

   f)

At–Home Help

If you have trouble visualizing the shape a net will make, copy the net onto grid paper, cut it out, and fold it.

2. Decide which net(s) will fold to make each structure.

   a)

   A   B   C

   b)

   A   B 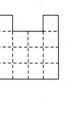   C

c)

A   B   C

d)

A   B   C

3. a) What shape do you think this net will make?

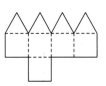

_____

b) Do a rough sketch of the shape in the space below.

4. In the space beside each structure, draw a net that could be used to build it.

a)

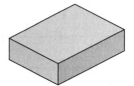

b)

c)

10.3 Top, Front, and Side Views of Cube Structures

▶ **GOAL: Recognize and sketch the top, back, front, and side views of cube structures.**

1. Draw the top, front, and side view of each structure. Use a thick black line to indicate a change in depth.

At-Home Help

You can use sugar cubes or blocks to build the structures in this lesson. Use grid paper to help you draw the structures. Remember that a thick black line shows a change in depth.

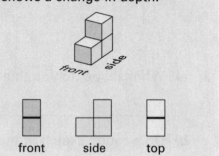

a)

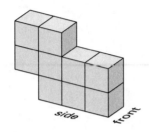

Top View	Front View	Side View

b)

Top View	Front View	Side View

c)

Top View	Front View	Side View

d)

Top View	Front View	Side View

10.4 Top, Front, and Side Views of 3-D Objects

▶ **GOAL: Recognize and sketch the top, front, and side views of 3-D objects.**

1. Identify the polyhedron that has these views. _____

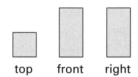

top   front   right

2. Identify the polyhedron that has these views. _____

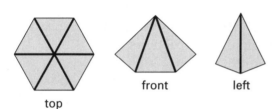

top   front   left

3. Draw the top, front, and side views of this polyhedron.

front

Top View	Front View	Side View

4. Draw the top, front, and side views of this structure.

Top View	Front View	Side View

5. Draw the top, front, and side views of this structure.

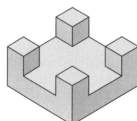

Top View	Front View	Side View

Isometric Drawings of Cube Structures

▶ **GOAL: Make realistic drawings of cube structures on triangle dot paper.**

1. The diagram below shows an isometric drawing of
   a 3-D letter H built out of cubes.

On the triangle dot grid, make isometric drawings for the following 3-D
letters built out of cubes. (If you wish, you can build the structures first
using sugar cubes or blocks.)

a) I

b) F

c) L

Isometric Drawings of 3-D Objects

▶ **GOAL: Make realistic drawings of 3-D objects on triangle dot paper.**

1. Use the triangle dot grid to make an isometric drawing of each structure.

a)

3 cm

3 cm 4 cm

At-Home Help

When drawing a 3-D object, first look for the basic shapes in the object. For example, if you are drawing a house, you can think of it as a cube with a triangular prism on top. You can draw these structures separately, and then put them together.

b)

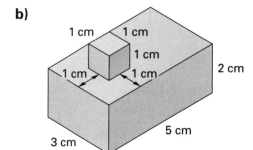

1 cm   1 cm
       1 cm
1 cm   1 cm   2 cm
       5 cm
3 cm

Communicating about Views

▶ **GOAL: Use mathematical language to describe views of 3-D objects.**

1. **a)** Write a paragraph that explains how to draw top, front, and side views of a pyramid with a square base. Use the Communication Checklist.

   _____

   _____

   _____

   _____

   _____

   b) Follow your directions to draw a top, front, and side view of the pyramid.

Top View	Front View	Side View

2. Write a paragraph that explains how to build a structure using the top, front, and side views shown below.

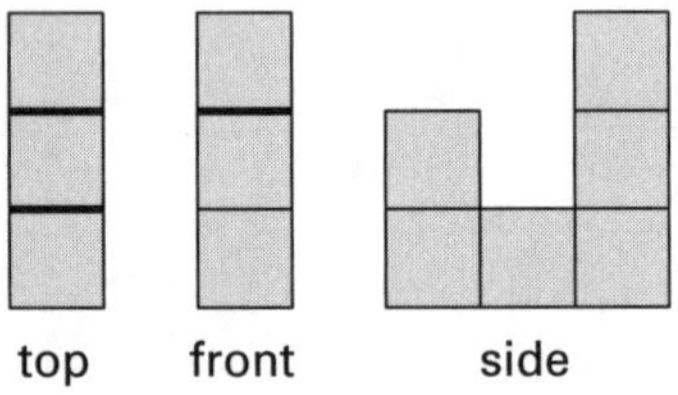

   top   front   side

   _____

   _____

   _____

   _____

   _____

   _____

   _____

Test Yourself

1. Which net will fold to form the polyhedron?

A   B   C

2. Which net(s) will fold to make this cube structure?

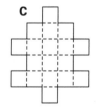

A   B

C

3. Draw a top, front, and right view of the structure. (You can build the structure from sugar cubes, blocks, or linking cubes to help you.)

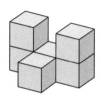

Top View	Front View	Side View

4. Make an isometric drawing of the cube structure. (You can build the structure from sugar cubes, blocks, or linking cubes to help you.)

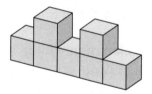

5. Make an isometric drawing of each structure on the triangle dot grid.

a)

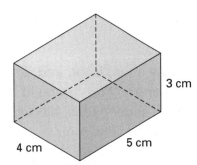

3 cm
4 cm 5 cm

b)

2 cm
2 cm
2 cm
2 cm
4 cm

6. Indira wrote a paragraph to explain how to draw top, front, and right views of the following structure.

Use the Communication Checklist to write an improved explanation of how to draw the views.

Indira's explanation: "First, I imagined that I was looking at the structure from above. There is one cube raised above the rest of the cubes. I used thick black lines to show the change in depth. Next, I imagined that I was looking at the structure from the front. There are two cubes on the bottom, and one cube on the upper left. Finally, I imagined I was looking at the structure from the right. There are three cubes on the bottom and one on the upper right."

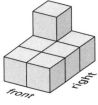

front   right

_____

_____

_____

_____

_____

_____

_____

_____

_____

▶ **GOAL: Develop a formula to calculate the surface area of a rectangular prism.**

1. A rectangular prism has the following dimensions: length = 4 cm, width = 2 cm, height = 6 cm.

   At-Home Help

   You can find the surface area of any prism by finding the area of each face and adding them together.

   For a rectangular prism, the front and back, top and bottom, and sides are all congruent. To find the surface area of a rectangular prism, you can find the areas of three different faces, add them together, and then double the sum. This can be written as a formula:
   Surface Area = 2 × (sum of the three different faces)

   a) What are the dimensions of the front face?

   Find the area of the front face.

   b) What are the dimensions of the top face?

   Find the area of the top face.

   c) What are the dimensions of one side face?

   Find the area of the side face.

   d) Add the three areas together and multiply by 2 to find the surface area of the rectangular prism.

2. Calculate the surface area of each prism.

   a)  **b)**   **c)**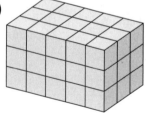

3. Calculate the surface area of each prism.

   a)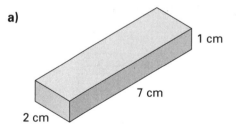
   1 cm
   7 cm
   2 cm

   b)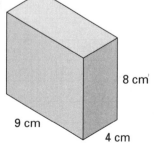
   8 cm
   9 cm
   4 cm

   c)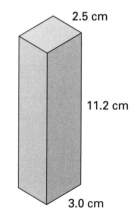
   2.5 cm
   11.2 cm
   3.0 cm

Volume of a Rectangular Prism

▶ **GOAL: Develop a formula to calculate the volume of a rectangular prism.**

1. Calculate the number of cubes in each prism.

   At-Home Help

   The formula to calculate the volume of a rectangular prism is
   Volume = length × width × height.

   a)

   b)

   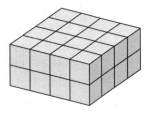

   c)

   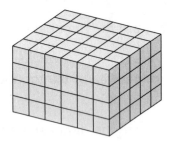

2. Calculate the volume of each prism.

   a)

   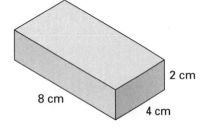

   b)

   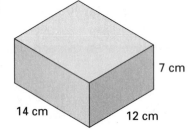

3. Fill in the table to calculate the volume of each rectangular prism. Use a calculator.

	Length (cm)	Width (cm)	Height (cm)	Volume (cm³)
a)	6	1	2	
b)	10	9	5	
c)	4	4	4	
d)	7	14	2	
e)	100	20	50	
f)	12.5	22.0	4.5	

4. Rana has bought a new fish tank with the following dimensions: length = 40 cm, width = 30 cm, height = 25 cm. She wants to buy a filter as well, to keep the water clean. Which filter should she buy?

   a) the Happy Fish filter, for tanks with volumes of 1000 to 5000 cm^3

   b) the Extra Pure filter, for tanks with volumes of 5000 to 20 000 cm^3

   c) the Super Clean filter, for tanks with volumes of 20 000 to 50 000 cm^3

5. Fill in the blanks in the table.

	Length (cm)	Width (cm)	Height (cm)	Volume (cm^3)
a)	3		1	12
b)	8	5		160
c)		3	3	81
d)	13	4	5	
e)	11		10	880
f)	24	15		4320

6. Simon's room has a volume of 60 m^3. What are two possible sets of dimensions (length, width, and height) for Simon's room?

   #1: _____ #2: _____

7. Sandra wants to buy a carton of juice. Three different brands are available:
   • Brand 1 costs $4 and has length = 5 cm, width = 5 cm, height = 15 cm
   • Brand 2 costs $2 and has length = 4 cm, width = 6 cm, height = 6 cm
   • Brand 3 costs $10 and has length = 10 cm, width = 8 cm, height = 15 cm

   Which brand is the best buy? Explain your answer.

8. Kwami folded this net into a box with a grey top and bottom, and white sides. Calculate the volume of the box in units cubed.

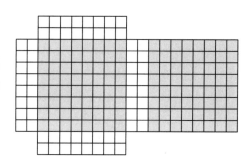

9. Jody offers Miguel a deal: she will trade her box of granola bars, which has the dimensions 5 cm by 6 cm by 7 cm, for his box, which has the dimensions 10 cm by 10 cm by 2 cm. Should he accept?

Solve Problems by Guessing and Testing

▶ **GOAL: Use guess and test and a table to solve measurement problems.**

1. Write 120 as the product of three numbers in five different ways. The first one is done for you.

   a) 120 = 3 x 4 x 10

   b) 120 = _____

   c) 120 = _____

   d) 120 = _____

   e) 120 = _____

2. Use your answers from question 1 to sketch five rectangular prisms, each with a volume of 120 cm³. Label all dimensions. The first one is done for you.

   a)

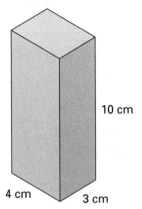

   10 cm
   4 cm
   3 cm

   d)

   b)

   e)

   c)

3. Sketch three rectangular prisms with a volume of 180 cm³. Label all the dimensions.

   a)   b)   c)

4. Paul is building a wooden chest. The volume of the chest will be 200 000 cm³, and the height will be 50 cm. Paul wants to use as little wood as possible to build the chest. What dimensions should he use for the base? Guess and test using the table below to find the answer.

Length (cm)	Width (cm)	Height (cm)	Surface Area (cm²)
100	40	50	
		50	
		50	
		50	
		50	
		50	
		50	

   Dimensions of the base: _____

5. A box has a volume of 7500 cm³ and a surface area of 2600 cm². What are the length, width, and height of the box?

11.4 Relating the Dimensions of a Rectangular Prism to Its Volume

▶ **GOAL:** Discover how changing the sides of a rectangular prism affects its volume; sketch a rectangular prism given its volume.

1. A rectangular prism has a volume of 240 cm^3.

   a) Write two possible sets of dimensions for the prism (length, width, and height).

   i) _____

   ii) _____

   At–Home Help

   Changing one dimension of a rectangular prism will result in a change to its volume. For example, if you double the height of a prism, its volume will also double.

   b) Look at the first set of dimensions you wrote in part (a). Predict the new volume you would get if you doubled the height. Explain your answer.

   _____

   _____

   c) Look at the second set of dimensions you wrote in part (a). If you doubled the height, would the new volume be equal to the volume in part (b)? Explain your answer.

   _____

   _____

   d) Double the height of each set of dimensions from part (a). Calculate the new volumes.

   i) _____ ii) _____

2. A rectangular prism has a height of 10 cm and a volume of 360 cm^3.

   a) The height of the prism is doubled. What is the new volume? _____

   b) The height of the prism is halved. What is the new volume? _____

   c) The height of the prism is tripled. What is the new volume? _____

   d) The height of the prism changes, making the volume 180 cm^3. The other dimensions stay the same. What is the new height? _____

   e) The height of the prism changes, making the volume 3600 cm^3. The other dimensions stay the same. What is the new height? _____

   f) The height of the prism changes, making the volume 36 cm^3. The other dimensions stay the same. What is the new height? _____

11.5 Exploring the Surface Area and Volume of Prisms

▶ **GOAL: Investigate relationships between surface area and volume of cubes and other rectangular prisms.**

1. Find the surface area and the volume of each rectangular prism.

a)

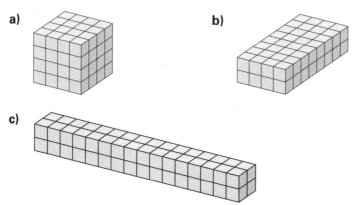

b)

c)

At-Home Help

To calculate the surface area of a rectangular prism, use this formula:

Surface Area = 2 × (Sum of three different faces)
= (2 × Area of top or bottom)
+ (2 × Area of front or back)
+ (2 × Area of one side)

To calculate the volume of a rectangular prism, use this formula:

Volume = length × width × height

2. Find the surface area and the volume of each rectangular prism.

a)

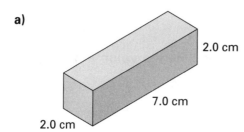

2.0 cm
7.0 cm
2.0 cm

b)

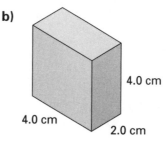

2.0 cm
4.0 cm
4.0 cm

c)

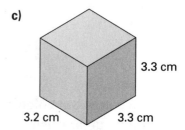

3.3 cm
3.2 cm   3.3 cm

3. Predict which prism has the greatest surface area. Explain your answer.

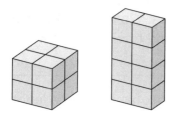

4. Predict which prism has the greatest volume. Explain your answer.

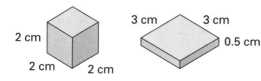

2 cm
2 cm   2 cm

3 cm   3 cm
0.5 cm

Test Yourself

1. Calculate the surface area of each rectangular prism.

   a)

   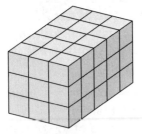

   b)

   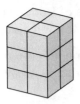

   c)

   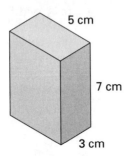

   5 cm
   7 cm
   3 cm

   c)

   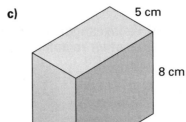

   5 cm
   8 cm
   9 cm

2. Calculate the volume of each rectangular prism.

   a)

   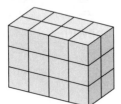

   b)

   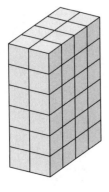

3. A box is 6 cm long, 4 cm wide, and 10 cm high.

   a) Calculate the surface area of the box.

   b) Calculate the volume of the box.

   c) A second box has the same length and width as the first box, but it is half as tall. What is the volume of the second box?

   _____

   d) A third box has the same length and height as the first box, but the width is doubled. What is the volume of the third box?

   _____

4. Fill in the spaces in the table below.

	Length	Width	Height	Volume	Sketch of prism
a)	2 cm	5 cm	3 cm		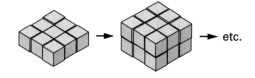
b)	4 cm	1 cm		4 cm³	
c)	3 cm	4 cm		36 cm³	
d)	5 cm	5 cm	5 cm		
e)	7 cm	1 cm	3 cm		

5. Sandra is building a tower out of blocks. She starts with a base that is 3 blocks long and 3 blocks wide. To make the tower higher, Sandra adds layers of 9 blocks at a time.

Sandra wants the surface area of her tower to have a number value that is double the number value of the volume. How tall should she make her tower?

12.1 Exploring Probability

▶ **GOAL: Determine probability from an experiment.**

1. Use the probability scale below to indicate the probability of each of the following.

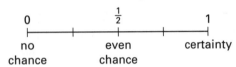

a) You will listen to the radio today. _____

b) If you drop a CD, it will land face up. _____

c) Your pet will live to be 250 years old. _____

d) If you toss a coin, it will land on its edge. _____

e) If you roll a regular six-sided die, it will show an even number. _____

At–Home Help

Probability is a number that shows how likely it is that an event will happen. For example, suppose you have a bag with two red marbles and three blue marbles. What is the probability of getting a red marble if you take one out without looking?

There are two red marbles out of a total of five. The probability is two out of five. You could write the probability as a fraction: $\frac{2}{5}$. You could also write it as a decimal: 0.4.

2. Omar, Kaitlyn, and Tynessa play a game with an ordinary six-sided die. If the die lands showing 1, 2, or 3, Omar wins. If it lands showing a 4 or 5, then Kaitlyn wins. If it lands showing a 6, then Tynessa wins.

a) Is this a fair game? Explain your answer.

_____

_____

_____

b) Who is the most likely to win? _____

3. A bag of marbles contains two red marbles, three blue, two yellow, and five green. Paul takes a marble without looking.

a) List all the possible outcomes.

_____

_____

b) What is the probability that he will get a blue marble?

_____

12.2 Calculating Probability

▶ **GOAL: Identify and state the theoretical probability of favourable outcomes.**

1. You have a red disc, a green disc, and a blue disc in a bag. You take out one disc without looking. Express each probability as a fraction.

   a) $P(\text{red}) =$ _____

   b) $P(\text{blue}) =$ _____

2. From the same bag as in question 1, you take out two discs without looking.

   a) What are the possible outcomes? _____

   _____

   b) What is $P(\text{one green and one blue})$? Express your answer as a decimal.

At-Home Help

A **favourable outcome** is the result that you are investigating in a probability experiment.

The **theoretical probability** is a measure of the likelihood of an event, based on calculations. You can calculate the theoretical probability using this ratio:

$$\frac{\text{number of favourable outcomes for an event}}{\text{total number of possible outcomes}}$$

The probability of an event is often written as $P(X)$, where X is a description of the event. For example, the probability of getting a plain chocolate out of a box of chocolates could be written as $P(\text{plain})$.

3. Indira plays on her school basketball team. In preparation for her next game, she practises making foul shots. She takes 30 practice throws and sinks 6 of them. Based on this information, what is the probability that she will make her next foul shot?

4. A gumball machine contains a mixture of 100 red, 100 blue, 100 green, and 100 yellow gumballs. The machine will give you only one gumball.

   a) What is the probability of getting a green gumball, $P(\text{green})$?

   b) What is the probability of getting $P(\text{other colour})$, which represents any colour other than green?

Solve Problems Using Organized Lists

▶ **GOAL: Use organized lists to determine all possible outcomes.**

1. Jody has three bills in her wallet. She left the wallet at home, and can't remember whether they are $5, $10, or $20 bills. Jody wants to figure out the probability of having $60 in her wallet, so she starts to make an organized list.

At-Home Help

Making a chart of all the possible outcomes can help you calculate probabilities. Work systematically and look for patterns in the chart to make sure you have not missed or repeated any of the outcomes.

$5 bills	3	2	2	1		1	0	0	0	0
$10 bills	0		0	2	0		3		2	1
$20 bills	0	0		0	2	1				
Sum	$15	$20			$45		$30	$60		

a) Fill in the blanks in Jody's chart.

b) How many different combinations are possible? _____

c) How many of these combinations add up to $60? _____

d) What is the probability that Jody has $60 in her wallet?

_____

e) What is the probability that Jody has $30 or more in her wallet?

_____

2. A football team can win, lose, or tie a game.

a) Use the chart below to find all the possible outcomes after four games.

Win																
Lose																
Tie																

b) Assume that all outcomes are equally likely. What is the probability that the team will win all four games?

12.4 Using Tree Diagrams to Calculate Probability

GOAL: Use tree diagrams to determine all possible outcomes.

1. In Yuki's kitchen, there are three kinds of bread: white, whole wheat, and pumpernickel. There are three kinds of jam: blueberry, strawberry, and raspberry. Yuki's sister gave her a piece of bread with jam.

   a) Use a tree diagram to show all the possible combinations.

   b) What is the probability that Yuki had blueberry jam on her bread?
   _____

   c) What is the probability that Yuki had pumpernickel bread with strawberry jam? _____

2. A family has five children.

   a) Use a tree diagram to show the different combinations of boys and girls in the family.

   b) What is the probability that four of the children will be boys?
   _____

   c) What is the probability that two of the children will be girls?
   _____

Applying Probabilities

▶ **GOAL: Calculate and compare probabilities.**

1. The girls' basketball team has been practising foul shots. Fill out the chart below to calculate the probability of each girl getting her next shot in the basket. Which girl is most likely to make her next shot?

Name	Try #1	Try #2	Try #3	Try #4	Try #5	Probability of making a shot
Indira	yes	no	yes	yes	no	
Tynessa	no	no	yes	no	yes	
Romona	yes	yes	yes	no	yes	
Kaitlyn	yes	no	yes	no	no	
Sandra	no	no	no	yes	no	

2. Bonnie, Indira, Chang, and Simon have been running races over the last two weeks. Bonnie won $\frac{1}{8}$ of the races. Indira won 50%, and Chang won one out of every four races. Simon's average was 0.125.

   a) Who is the most likely to win the next race?

   b) Which two runners have the same probability of winning the next race?

3. Paul and Colin are playing a game. Paul tosses two coins, and wins if he gets two heads. Colin rolls one die, and wins if he gets a 6.

   a) What is the probability of Paul winning?

b) What is the probability of Colin winning?

c) Who is the most likely to win the game?

4. A knapsack holds three textbooks of the same size: a math book, a science book, and a history book. If Jody reaches in and grabs a book, what is the probability that she will get the math book?

5. Bag A has five red marbles and two green ones. Bag B has seven red marbles and three green ones.

   a) Which bag has the greater probability of drawing a green marble?

   b) If all the marbles are placed in one bag, what is the probability of drawing a red marble?

6. James throws a pair of dice and gets a total of 2. Chang tosses three coins, and gets one tails and two heads. Which event was the least likely?

7. What total are you most likely to get when throwing a pair of dice? Explain your answer.

   _____

   _____

8. Two out of every three trees in Sunnyside Park are Norway maples. During a game of tag, Sandra touches three trees. Is it certain that she has touched at least one Norway maple? Explain your answer.

   _____

   _____

Test Yourself

1. A bag has three marbles in it. Two are white and one is black.

   a) If you reach in and take one marble out, what is the probability of getting a white marble?

   _____

   b) If you take a white marble out, and then reach in again, what is the probability of getting a white marble this time? _____

2. Another bag has 10 marbles in it: 2 yellow, 4 green, 3 black, and 1 red. Romona reaches in and takes one out. What is each probability?

   a) P(black)

   b) P(yellow)

   c) P(red)

   d) P(green)

3. Yuki has four coins in her hand. She says they are worth 40¢, and there are no pennies.

   a) Fill out the organized chart at the bottom of the page to list all the possible combinations.

   b) If you take one guess, what is the probability that you can guess what coins are in Yuki's hand?

   c) If the coins in Yuki's hand are worth 70¢, what is the probability that you can guess what they are in one guess?

4. A small book contains three short stories. Each story is at least one page long. The total number of pages is seven.

   a) Use an organized chart to list all the different combinations of pages the stories could have.

   b) What is the probability that one of the stories is five pages long?

   c) What is the probability that one of the stories is four pages long?

Quarters	4	3	3	2	2	2	1	1	1	1	0	0	0	0	0
Dimes															
Nickels															
Total value															

5. In a random maze, each intersection gives you an option: right or left. On the path through the maze, there are four intersections (A, B, C, and D) before reaching the centre.

a) Use a tree chart to show all the possible turnings the path could take.

b) What is the probability that the path to the centre turns right four times?

c) What is the probability that the path to the centre turns left twice and right twice, in any order?

6. Indira has two pairs of shoes, two pairs of jeans, and three sweaters. Use a tree diagram to show all the outfits she can put together.

7. Four soccer teams have the following records:
   • Team 1 won 5 out of the last 20 games.
   • Team 2 wins 40% of the games they play.
   • The likelihood of Team 3 winning its next game is $\frac{1}{3}$.
   • Team 4 won 3 out of the last 8 games.

a) Which team is the most likely to win?

b) Which team is the most likely to lose?

8. Colin's wallet has one $20 bill, three $10 bills, and two $5 bills. Tynessa's wallet has five $10 bills and two $20 bills. Simon's wallet has three $20 bills and nine $5 bills.

a) Romona is allowed to take one bill out of one wallet without looking. Which wallet should she choose to get the greatest chance of a $20 bill?

b) All the bills are put together in a bag. What is the probability of drawing a $20 bill?

Answers

Chapter 1

1.1 Using Multiples

1. a) 10, 20   b) 10
2. a) 3, 6, 9, 12, 15, 18, 21, 24, 27, 30
   b) 4, 8, 12, 16, 20, 24, 28, 32, 36, 40
   c) 7, 14, 21, 28, 35, 42, 49, 56, 63, 70
3. 84
4. a) 4   b) 30   c) 40   d) 12
   e) 24   f) 30
5. once
6. 20 days

1.2 A Factoring Experiment

1. a) There is a diagonal pattern running down and to the left across the chart. Below each number that has a factor of 3, there are two numbers that do not.
   c) 1, 9, and 27
2. a) In every other column, every other number is highlighted.
   c) 1, 2, 8, 16, 32
3. a) 12, 36, 48, 60, 72, 84, and 96 all have both 3 and 4 as factors.
   b) $1 \times 24 = 24$, $2 \times 12 = 24$, $3 \times 8 = 24$, $4 \times 6 = 24$, $6 \times 4 = 24$, $8 \times 3 = 24$, $12 \times 2 = 24$, $24 \times 1 = 24$
   c) 1, 2, 3, 4, 6, 8, 12, 24

1.3 Factoring

1. a) $12 \div 1 = 12$, $12 \div 2 = 6$, $12 \div 3 = 4$, $12 \div 4 = 3$, $12 \div 6 = 2$, $12 \div 12 = 1$
   b) When you divide 12 by 5, 7, 8, 9, 10, or 11, you do not get a whole number.
2. a) The missing factors are 3 and 9.
   b) The factors of 45 are 1, 3, 5, 9, 15, 45.
   c) The factors of 50 are 1, 2, 5, 10, 25, 50.
3. a) 1, 3, 5, 15
   b) 1, 2, 4, 5, 8, 10, 20, 40
   c) 1, 2, 3, 6, 9, 18, 27, 54
   d) 1, 2, 3, 4, 6, 8, 9, 12, 18, 24, 36, 72
4. a) 5   b) 8
5. a) 28   b) 13   c) 15

6. The books can be stacked into equal piles of 1, 7, 11, or 77.

1.4 Exploring Divisibility

1. a) no   b) yes   c) no   d) yes
2. a) yes   b) no   c) no   d) yes
3. a) no   b) yes   c) yes   d) yes
4. a) yes   b) no   c) yes
5. The chart forms the word "HI."

1.5 Powers

1. a) $3^3 = 27$ b) $6^2 = 36$ c) $2^6 = 64$
   d) $10^5 = 100\ 000$
2. a) $10 \times 10 \times 10 \times 10 = 10\ 000$
   b) $5 \times 5 \times 5 \times 5 \times 5 = 3125$
   c) $2 \times 2 \times 2 \times 2 \times 2 \times 2 \times 2 \times 2 = 256$
   d) $1 \times 1 \times 1 \times 1 \times 1 \times 1 \times 1 = 1$
3. a) true   b) false: $3^3 = 3 \times 3 \times 3$
   c) true   d) false: $6^3 = 216$
4. a) 5^2 b) 4^3 c) 2^3 d) 3^3
5. a) 2^4 b) 16
6.

Power	Base	Exponent	Meaning	Product
3^2	3	2	3×3	9
2^3	2	3	$2 \times 2 \times 2$	8
6^3	6	3	$6 \times 6 \times 6$	216
4^2	4	2	4×4	16
3^4	3	4	$3 \times 3 \times 3 \times 3$	81

1.6 Square Roots

1. b) 9 square units; 3 units; 3
   c) 16 square units; 4 units; 4
   d) 25 square units; 5 units; 5
2. a) 6   b) 8   c) 7
3. 18 m by 18 m
4. a) 21   b) 22   c) 25   d) 32
   e) 82   f) 41

1.7 Order of Operations

1. a) 16   b) 40   c) 11   d) 4
   e) 9   f) 3   g) 10   h) 14
   i) 21   j) 97   k) 39   l) 13

2. a) $(3 + 2) \times 4 = 20$
 b) $20 - 6 \times (2 + 1) = 2$
 c) $4 \times (2 + 3) + 1 = 21$
 d) $4 \times (2 + 3 + 1) = 24$
 e) $(4 + 2) \times (3 - 2) = 6$
 f) $(7 - 2)^2 + 3 = 28$
 g) $(2 + 2 + 2)^2 = 36$
 h) $8 - 3 \times (2 - 1) = 5$
3. a) 21 **b)** 7 **c)** 2 **d)** 17
 e) 5 **f)** 38 **g)** 35 **h)** 8
 i) 23 **j)** 25
4. Fawn is correct.
5. (d)
6. (b)
7. a) < **b)** > **c)** > **d)** <
 e) = **f)** = **g)** < **h)** =
8. a) $3 \times 4 + 2 \times 6$
 b) 24 articles of clothing

1.8 Solve Problems by Using Power Patterns

1. a) The exponent is the same as the number of zeros.
 b) 100 zeros
2. 5
3. a) 16
 b) Square the middle number to find the sum. $5^2 = 25$
 c) 81

Test Yourself

1. a) 2, 4, 6, 8, 10, 12
 b) 8, 16, 24, 32, 40, 48
 c) 10, 20, 30, 40, 50, 60
 d) 12, 24, 36, 48, 60, 72
2. a) yes **b)** no **c)** yes **d)** yes
3. a) 30 **b)** 28 **c)** 24 **d)** 20
4. every 21 days
5. a) 1, 2, 3, 4, 6, 9, 12, 18, 36
 b) 1, 2, 3, 6, 7, 14, 21, 42
 c) 1, 2, 4, 17, 34, 68
 d) 1, 3, 9, 27, 81
6. a) 8 **b)** 1 **c)** 24 **d)** 14
7. a) 2 **b)** 7 **c)** 21
8. 2 buses, 1 van
9. a) Mr. Singh's garden could be 1 by 36, 2 by 18, 3 by 12, 4 by 9, or 6 by 6. Mrs. Jackson's garden could be 1 by 20, 2 by 10, or 4 by 5.
 b) The fence could be 4 m long (the GCF).
10. a) yes **b)** yes **c)** no **d)** no
 e) yes **f)** no
11. a) 32 **b)** 125 **c)** 10 000
12. a) 3^3 **b)** 2^7 **c)** 10^6
13. 17 m by 17 m

14. a) 10 **b)** 100 **c)** 200 **d)** 3000
15. a) 30 **b)** 13 **c)** 21 **d)** 4505
16. a) 1 **b)** 8 **c)** 5 **d)** 4
17. a) Volume = 5 cm³ **b)** 125 cm³

Chapter 2

2.1 Exploring Ratio Relationships

1. a) yes **b)** no
2. b) $1:2$, $3:6$, $4:8$; They are all similar.
3. c) yes **d)** $5:1$ **e)** $10:2$ **f)** yes

2.2 Solving Ratio Problems

1. a) $4:2$, $6:3$, $8:4$
 b) $8:1$, $16:2$, $24:3$
 c) $9:5$, $18:10$, $36:20$
 d) $\dfrac{20}{6}$, $\dfrac{30}{9}$, $\dfrac{40}{12}$
2. a) 2 **b)** 6 **c)** 6 **d)** 3
3. a) 16 shaded squares, $4:6 = 16:24$
 b) 2 shaded squares, $1:4 = 2:8$
4. a) $2:3$ **b)** 15 pails of earth
5. a) $12:6$ **b)** 36 years old
 c) three times

2.3 Solving Rate Problems

1. a) 9 times/3 months; 3 times/month; 6 times/2 months
 b) $32/4 h; $8/h; $16/2 h
2. a) 14 km/h **b)** 2 h
3. a) $18 **b)** $108
4. a) 36 times/min **b)** 72 times/min
 c) 130 times/min
5. a) 5 min **b)** 4 min

2.4 Communicating about Ratio and Rate Problems

1. yes
2. a) 4 km **b)** no, only 8 km
3. a) 20 brownies **b)** 25 brownies

2.5 Ratios as Percents

1. a) $\dfrac{1}{2}$ or $\dfrac{50}{100}$; 0.50; 50%

 b) $\dfrac{3}{4}$ or $\dfrac{75}{100}$; 0.75; 75%

 c) $\dfrac{1}{2}$ or $\dfrac{50}{100}$; 0.50; 50%

 d) $\dfrac{1}{2}$ or $\dfrac{50}{100}$; 0.50; 50%

e) $\frac{28}{100}$ or $\frac{7}{25}$; 0.28; 28%

f) $\frac{40}{100}$ or $\frac{2}{5}$; 0.40; 40%

2. a) 46%   b) 38%   c) 48%   d) 40%
   e) 80%   f) 75%

3. a) 0.21, 0.74, 0.98, 0.03
   b) 21 : 100, 74 : 100, 98 : 100, 3 : 100

4. a) 1 : 4   b) $\frac{3}{5}$ c) 1 : 5, 0.2
   d) $\frac{3}{10}$, 30% e) 0.41, 41%
   f) 9 : 50, 18%
   g) $\frac{23}{100}$, 23 : 100, 0.23   h) $\frac{6}{8}$, 0.75, 75%

5. a) 100%, 84%, 64%, 41%, 2%
   b) 114%, 35%, 14%, 7%, 3%
   c) 0.9, 74%, 0.5, 0.32, 19%
   d) 88%, 56%, $\frac{45}{100}$, $\frac{2}{100}$
   e) 85%, 0.81, 44%, $\frac{1}{4}$, $\frac{2}{10}$
   f) 91%, 72%, $\frac{8}{20}$, $\frac{26}{100}$, 0.04

6. 25%
7. 20%
8. 15%

2.6 Solving Percent Problems

1. a) 30   b) 10   c) 6   d) 3.6
   e) 67.5   f) 12
2. a) 300   b) 150   c) 70   d) 750
   e) 800   f) 40
3. 28%
4. 106%
5. a) $16.08, $18.00, $17.25
   b) Able Audio
6. 40, 64, 16
7. 585 people
8. a) 12 matches   b) 4 matches
9. 30%
10. a) $3.00, $27.00   b) $4.95, $10.05
    c) $22.50, $22.50   d) $2.50, $47.50
    e) $14.70, $83.30   f) $16.50, $5.50
11. a) 70%   b) $175   c) $52.50

2.7 Decimal Multiplication

1. a) 0.15   b) 0.54
2. a) 0.14   b) 0.48   c) 1.55   d) 0.096
3. 41.25 km
4. $3.15

2.8 Decimal Division

1. a) 13   b) 12
2. a) 4   b) 33   c) 4.16   d) 9.2
3. a) 30 flags b) 5 flags c) 16 flags
4. 12.5 h

Test Yourself

1. a) 40 : 60   b) $\frac{2}{5}$ c) 60%
2. (a), (c), and (d) are equivalent
3. a) 4   b) 6   c) 15   d) 4
   e) 1   f) 8   g) 3   h) 30
4. 8 min
5. $12/h, $8.50/h, $10/h, $7.75/h
6. 1350 people
7. a) 110   b) 10   c) 60   d) 20%
   e) 140   f) 22
8. a) 25%   b) 75 dogs
9. a) 5.28   b) 0.015   c) 1.98   d) 4
   e) 5   f) 3.90
10. 0.96 (Romona), 0.87 (Miguel), 0.79 (Paul), 0.75 (Fawn)
11. 336 people
12. a) 180 cm by 90 cm   b) 150 cm
    c) 150 cm by 240 cm   d) 210 cm

Chapter 3

3.1 Collecting Data

1. primary data
2. a) 15% chose pizza, 50% chose hot dogs, 30% chose hamburgers, 5% chose sandwiches
   b) hot dogs and hamburgers
3. a) 70% chose apple juice, 25% chose grape juice, 5% chose orange juice
   b) apple juice

3.2 Avoiding Bias in Data Collection

1. a) ii   b) ii   c) i
2. a) What is your favourite type of cereal? Or, which of the following cereals do you prefer? (with a list provided)
   b) What type of music do you listen to?

3.3 Using a Database

1. a) a field   b) a record
2. Number in store
3. a) Sale price   b) $20 for jeans
   c) $75 for black sweater

4. the black sweater
5. original price

3.4 Using a Spreadsheet

1. a) $49.99   **b)** 25
2. a) $12.95
   b) the number of items
   c) the record for the shorts
3. $9.95 × 20 = $199.00
4. B3*C3 and B4*C4
5. =sum(D2:D5)

3.5 Frequency Tables and Stem-and-Leaf Plots

1. a) 35 people   **b)** January
2. a) 99   **b)** 51   **c)** 16 people
   d) 10 people
3.

Examination Mark	
Interval	**Frequency**
0–10	0
11–20	0
21–30	0
31–40	0
41–50	0
51–60	3
61–70	4
71–80	6
81–90	8
91–100	6

4. a)

Time (min)	
Stem	**Leaf**
1	4 6 9
2	2 5 8
3	1 3 5 8
4	0 4 5 8 9
5	0 0 2 5
6	2 4 6 7
7	3 4 5 6
8	1 8
9	3 6
10	2

   b) 32 students   **c)** 14 min   **d)** 9%

3.6 Mean, Median, and Mode

1. a) 7.3, 7.5, 9   **b)** 14.3, 14.5, 15
   c) 55.1, 54, 54   **d)** 4.4, 4, 4
   e) 35.7, 26, 23   **f)** 51.75, 57, 2 and 56

2. a) $892.86; $7.50; $7.50   **b)** the mean
   c) the mean
3. a)

Type of juice	Frequency
apple	12
orange	32
lemonade	15
grape	23
grapefruit	4

   b) 17.2, 32, no mode   **c)** orange   **d)** grapefruit

3.7 Communicating about Graphs

1. a) sour cream   **b)** plain
   c) sour cream, BBQ, salt & vinegar
2. The cafeteria should order more salad, less soup, and the same amount of pizza for the next month.

Test Yourself

1. a) the book club, and maybe the students also
   b) families in the neighbourhood
2. a) Teriyaki stir-fry
   b) Number of people it serves, or Price
3. For example, "How many hours of TV do you watch every week?"
4. a) 6   **b)** sum(B2:D2)   **c)** 14
5. b) 50.1; 48; 46
6. a) 9 h   **b)** 6.25%

Chapter 4

4.1 Exploring Number Patterns

1. a) 34, 32, 30, 28, 26   **b)** 16, 26, 36, 46, 56
   c) 6, 12, 24, 48, 96   **d)** 100, 10, 1, 0.1, 0.01
2. a) Add the two numbers above each box to get the number in the box.
   b) The missing numbers are 67, 103, and 170.
3. a) The next arrow will point down. The arrow after that will point down and to the left.
   b) The next figure will have 9 squares at the bottom and be 5 squares high. The figure after that will have 11 squares at the bottom and be 6 squares high.

4.2 Applying Pattern Rules

1. a) 15, 18, 21; Rule: Add 3 to each number to get the next.
   b) 21, 25, 29; Rule: Add 4 to each number to get the next.

c) 256, 1024, 4096; Rule: Multiply each number by 4 to get the next.

d) 64, 55, 46; Rule: Subtract 9 from each number to get the next.

e) 160, 320, 640; Rule: Multiply each number by 2 to get the next.

f) 80, 40, 20; Rule: Divide each number by 2 to get the next.

g) 2592, 15 552, 93 312; Rule: Multiply each number by 6 to get the next.

h) 65, 129, 257; Rule: Multiply each number by 2 and subtract 1 to get the next.

2. 1, 5, 25, 125, 625, …

3. 4, 7, 13, 25, 49, …

4. a) 25, 30, 35; Rule: Add 5 to each number.

b) 100, 10, 1; Rule: Divide each number by 10.

c) 1, 10, 100; Rule: Multiply each number by 10.

d) 0.0625, 0.03125, 0.015625; Rule: Divide each number by 2.

e) 625, 3125, 15 625; Rule: Multiply each number by 5.

f) 289, 278, 267; Rule: Subtract 11 from each number.

g) 1.8, 2.2, 2.6; Rule: Add 0.4 to each number.

h) 27, 9, 3; Rule: Divide each number by 3.

5. 2, 3, 5, 9, 17, 33, 65, 129, …

6. All the numbers in the sequence are 1.

7. a) Multiply by 1, then 2, then 3, and so on.

b) 720, 5040, 40 320

4.3 Using a Table of Values to Represent a Sequence

1. a) The next two values are 17 and 20.

b) Add 3 to each value to get the next.

c) Multiply the term number by 3 and add 2.

d) 26

2. a) 35, 42, 49, 56

b) Multiply the term number by 7.

c) 140

3. a) The missing numbers are 26, 31, 36, and 41.

b) Multiply the term number by 5 and add 1.

c) 81

4. b) The missing values are 3, 5, 7, and 9.

c) Add 2 to each value to get the next; or multiply each term number by 2 and add 1.

d) 21 toothpicks

4.4 Solve Problems Using a Table of Values

1. The missing values are 3, 8, 13, 18, 23, 28, 33, and 38. The 15th figure will have 73

boxes. (Multiply the term number by 5 and subtract 2.)

2. a) 2 cards

b) 6 cards

c) 12 cards

d) The missing values are 0, 2, 6, 12, 20, 30, and 42.

e) Pattern rule: Add 2, 4, 6, and so on, to each number to get the next number.

f) 42 cards

3. 45 games

4. the seventh day of work

5. 5 weeks

6. a) 195 bars   b) on the 13th day

7. 8 days

4.5 Using a Scatter Plot to Represent a Sequence

1. The missing values are 3, 11, and 19.

2. a) 6 posts, 10 rails   b) 12 posts, 22 rails

3. a) 5 links

b)

Term number (chain number)	Term value (number of links)
1	5
2	9
3	13
4	17

c) 37 links

Test Yourself

1. a) 10, 12, 14; Rule: +2

b) 15, 21, 28; Rule: +1, +2, +3, …

c) 256, 1024, 4096; Rule: ×4

d) 50, 98, 194; Rule: ×2 then −2

2. a)

Term number	Term value (number of circles)
1	1
2	3
3	6
4	10
5	15
6	21
7	28

b) The seventh figure has 28 circles in it.

c) Pattern rule: Add 2, 3, 4, and so on, to each number to get the next number.

3. a) 0, 5, 10, 15, 20, 25

b) 2, 12, 72, 432, 2592, 15 552

c) 100, 60, 40, 30, 25, 22.5

d) 1, 2, 5, 14, 41, 122

4. 5 days

5. b) the sixth figure

   c) 20 white squares, 16 shaded squares

6. b) 16 circles

Chapter 5

5.1 Area of a Parallelogram

1. a) 4 units   **b)** 6 units   **c)** 24 units squared

2. a) 15 cm^2 **b)** 8 m   **c)** 5 cm   **d)** 16.96 m^2

   e) 1.5 mm **f)** 0.5 dm

3. A: 6 units squared B: 18 units squared

   C: 20 units squared

5.2 Area of a Triangle

1. a) 24 m^2 **b)** 14 cm^2

2. a) 36 cm^2 **b)** 8 mm   **c)** 20 m   **d)** 87.3 cm^2

3. 360 cm^2

4. a) 6 cm^2 **b)** 12 cm^2 **c)** 6 cm^2 **d)** 24 cm^2

5.3 Calculating the Area of a Triangle

1. a) 3 units squared   **b)** 3 units squared

   c) 6 units squared

2. Your triangles could have $h = 4$, $b = 12$;

   $h = 6$, $b = 8$; $h = 2$, $b = 24$; $h = 8$, $b = 6$;

   $h = 12$, $b = 4$; or $h = 24$, $b = 2$.

3. a) 20 m^2

   b) The height of the second triangle is 8 m, while the height of the first triangle is 10 m. The bases are the same. So the second triangle should have a smaller area than the first triangle.

   c) 16 m^2

   d) To find the area, you will multiply the base by the height and divide by 2. So the calculation will be the same whether $b = 4$ and $h = 10$ or $b = 10$ and $h = 4$. The two triangles will have the same area.

4. a) 8000 cm^2 or 0.8 m^2

   b) 2000 cm^2 or 0.2 m^2

   c) Although the base and height of the triangles are fixed, your triangles can be various shapes such as symmetrical, slanted to the left, or slanted to the right.

5.4 Area of a Trapezoid

1. a) 20 units squared   **b)** 36 units squared

2. 280 cm^2

3. 6 m

4. Your trapezoid could have sides of 2, 4, and $h = 3$, or sides of 4, 5 and $h = 2$, among other solutions.

5.5 Exploring the Area and Perimeter of a Trapezoid

1.

	Side length (cm)	Side length (cm)	Base a (cm)	Base b (cm)	Height h (cm)
Trapezoid A	3.5	3.5	3	2	3.4
Trapezoid B	2.5	2.5	4.5	2.5	2.4
Trapezoid C	1	1	5.5	4.5	0.8

2. a) 12 cm

   b) Trapezoid A will probably have the greatest area. It looks the largest and is the closest in shape to a square, having the sides similar in length to the bases.

3. a) The three areas are 8.5 cm^2, 8.4 cm^2, and 4.0 cm^2.

   b) Trapezoid A has the greatest area.

5.6 Calculating the Area of a Complex Shape

1.

Area of rectangle	Area of triangle	Area of parallelogram	Area of trapezoid
28 m^2	10 m^2	15 m^2	15 m^2

Total area = 68 m^2

2. a) 39 cm^2 **b)** 52 cm^2

3. a) 42 m^2 **b)** 6 m

4. a) 47 m^2 **b)** \$376

5. a) 198 cm^2 **b)** 31.5 m^2 **c)** 8.25 m^2

   d) 318 cm^2

5.7 Communicating about Measurement

1. 120 cm, 684 cm^2

2. a) 52 cm^2; subtract the area of the parallelogram from the area of the square

   b) 273.75 cm^2

3. 0.4 m^2

Test Yourself

1. a) 15 cm^2 **b)** 24.5 cm^2 **c)** 13.86 cm^2

   d) 21.3 m^2

2. first triangle: 36 m^2, second triangle: 12 m^2

3. a) 2 m^2 **b)** 3.24 m^2 **c)** 32 cm^2

   d) 102.24 cm^2 **e)** 625 cm^2

4. 108 cm^2

5. a) 22 cm^2; find the area of the triangle

b) 42 m²; use the height of the rectangle as the height of the triangle

c) 85 cm²; either subtract the area of the central triangle from the area of the trapezoid, or find the area of the two other triangles separately and add them together

6. a) 43 cm² **b)** 31.25 m²

Chapter 6

6.1 Comparing Positive and Negative Numbers

1. −8, −7, −5, −4, −2, −1, 0, +1, +3, +5, +7, +8

2. a) −4, −3, 0, +3, +4
b) −6, −4, −2, +5, +9
c) −98, −6, +1, +22, +35
d) −67, −38, 0, +8, +45
e) −123, −8, +3, +46, +98

3. a) +1 **b)** −2 **c)** −6 **d)** 0
e) −1 **f)** +4 **g)** −1, 0, +1
h) 0

4. a) > **b)** < **c)** > **d)** <
e) > **f)** > **g)** < **h)** >

6.2 An Integer Experiment

1. Ellie is on floor 23.
2. POSITIVELY

6.3 Adding Integers Using the Zero Principle

1. a) +7 **b)** −6 **c)** +8 **d)** −11
2. a) +1 **b)** +2
3. a) −2 **b)** +3 **c)** −2 **d)** +5
4. $3
5. The shaded spaces will show the signs + / −.

6.4 Adding Integers That Are Far from Zero

1. a) (−3) + (−3) = (−6)
b) (−3) + (+2) = (−1)
c) (+2) + (−2) = 0
d) (+3) + (−3) = (0)
e) (+3) + (−1) = (+2)
f) (−5) + (+6) = (+1)

2. a) −8 **b)** +7 **c)** +1 **d)** −1
e) +7 **f)** −7

3. a) −15 **b)** −75 **c)** +75 **d)** +15
e) −125 **f)** −75 **g)** +125 **h)** +75
i) −34 **j)** −50 **k)** −20 **l)** −75

6.5 Integer Addition Strategies

1. a) +54 **b)** +92 **c)** −77 **d)** −56
e) +62 **f)** +387 **g)** −8

2. a) −6 **b)** −5 **c)** −25 **d)** +43
e) −6 **f)** −36
3. a) −70 **b)** −70 **c)** −88 **d)** −50
e) −82 **f)** +55
4. a) −10 **b)** +100 **c)** +34 **d)** −50

6.6 Using Counters to Subtract Integers

1. a) −4 **b)** −37 **c)** +6
2. a) +4 **b)** +7 **c)** +8 **d)** +50
e) −6 **f)** +2
3. a) −8 **b)** −1 **c)** −1 **d)** +6
4. ADD THE OPPOSITE.

6.7 Using Number Lines to Subtract Integers

1. b) (−11) − (+7) = (−18)
c) (−14) − (−26) = (+12)
2. a) −36 **b)** −28 **c)** +28 **d)** +36
e) +12 **f)** −7

6.8 Solve Problems by Working Backwards

1. a) −9 **b)** −12
2. The elevator started on floor 14.
3. Meagan started with $24.75.
4. Miguel started at the 4 m level and Yoshi started at the 6 m level.
5. Shailini must leave the house by 11:15 A.M.

Test Yourself

1. a) −4 **b)** +2 **c)** +3 **d)** +7
e) +2 **f)** +13
2. The order of integers on the number lines will be:
a) −5, −3, −1, 0, +5
b) −20, −15, −5, +10, +20
c) −13, −7, −5, +4, +9
3. −17, −5, −4, 0, +1, +2, +8, +17
4. a) < **b)** > **c)** > **d)** >
e) = **f)** = **g)** > **h)** <
i) < **j)** =
5. a) −7, −10, −13 **b)** +5, +9, +13
c) +6, −7, +8 **d)** 0, +3, +1
6. a) +8 **b)** −15 **c)** +4 **d)** 0
e) −15 **f)** +40 **g)** −40 **h)** +15
i) +98 **j)** +15
7. a) −24 **b)** −3 **c)** −10 **d)** +10
e) −31 **f)** +140 **g)** +6 **h)** +65
i) −102 **j)** +51
8. a) + **b)** − **c)** − **d)** +
9. a) +8 **b)** −7 **c)** −7 **d)** 11
e) 3 **f)** 51
10. (+8) − (−14) = (+22)
11. (−10) + (+14) = (+4)
12. (−3) − (+19) = (−22)
13. a) −2 **b)** −11

14. 24 cookies

15. $(+45) + (+25) + (+15) + (-40) = (+45)$

Chapter 7

7.1 Comparing Positions on a Grid

1. $B(2, 2)$, $C(-2, 3)$, $D(2, 0)$, $E(0, -3)$, $F(-1, -2)$

2. $A(7, -2)$, $B(3, 5)$, $C(-6, 8)$, $D(-4, 0)$, $E(0, 9)$, $F(-7, -5)$

3.

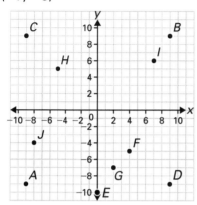

7.2 Translations

1. a) $A'(-2, 2)$ **b)** $B'(4, 5)$ **c)** $C'(-1, -1)$

2. $A'(2, -5)$, $B'(2, -1)$, $C'(6, -1)$, $D'(6, -5)$

3. $D'(2, -1)$, $E'(4, 2)$, $F'(6, 0)$

4. a) 4 units to the right
    b) 2 units down and 2 units to the right

7.3 Reflections

1. $A'(1, 5)$, $B'(-3, 5)$, $C'(-3, 2)$, $D'(1, 2)$

2. a)

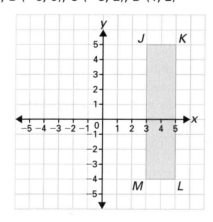

b)
    $J'(3, -5)$, $K'(5, -5)$, $L'(5, 4)$, $M'(3, 4)$
   c) $J''(-3, 5)$, $K''(-5, 5)$, $L''(-5, -4)$, $M''(-3, -4)$

3. $Q'(2, 4)$, $R'(3, 1)$, $S'(5, 3)$

7.4 Rotations

1. a) A   **b)** 90°   **c)** ccw

2. a)

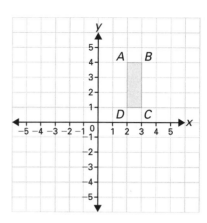

b) $A'(4, -2)$, $B'(4, -3)$, $C'(1, -3)$, $D'(1, -2)$

c) $A''(-4, 2)$, $B''(-4, 3)$, $C''(-1, 3)$, $D''(-1, 2)$

d) $A'''(-2, -4)$, $B'''(-3, -4)$, $C'''(-3, -1)$, $D'''(-2, -1)$

7.5 Congruence and Similarity

1. The shapes are congruent because they have the same shape and size.

2. a) A and C   **b)** D and F

3. The second figures should have the same size and shape.

4. The second figures should have the same shape but a different size.

5. b) The matching sides are different, but the matching angles are the same.

6. b) The matching sides and the matching angles are the same.

7.6 Tessellations

1. a) There are two different orientations.
   b) One orientation looks like a Z. The other orientation looks like a chair.
   c) To transform the Z orientation into the chair orientation, you need to do a 90° cw rotation.

7.7 Communicating about Geometric Patterns

1. a) (1) Draw a right triangle with a 2 unit base and a 4 unit height. The right angle should be at the bottom left vertex. The triangle should point up. (2) On the same base, draw a second triangle that is similar to the first, but has a 1 unit base and a 2 unit height. Make the second triangle oriented the same way as the first triangle. (3) Line up the base of the second triangle with the base of the first triangle. (4) Translate this triangle to the right until its right angle is 2 units to the right of the bottom right units of the first triangle.

b) (1) Draw a vertical line that is 4 units long. From the bottom of the line, draw a second line to an imaginary point 3 units to the right and 1 unit up. From the end of the second line, draw a third line to an imaginary point 3 units up and 3 units to the left. The three lines will form a triangle. (2) Rotate the triangle 90° ccw around the top vertex. Repeat two more times.

2. a) (1) Draw a pentomino that has four blocks in a straight horizontal line, and one block sticking up at the end. Shade the pentomino grey. (2) Reflect the pentomino in its base. Leave the image white.
(3) Reflect the first pentomino in a vertical line that touches the right-hand side of the pentomino. Leave the image white.
(4) Reflect the white pentomino on the right-hand side in its base. Shade the image.

b) (1) Draw a small white square, about 1 cm by 1 cm. (2) Draw four congruent isosceles triangles around the square. Each side of the square will be the base for one triangle. The height of each triangle will be double the base. (3) Draw a larger square around the figure. The tip of each triangle should touch the centre of one side of the larger square. Make the line around the larger square thicker than the rest of the lines in the figure.

7.8 Investigating Pattern Blocks

1. B, C, D
2. 90°
3. a) No. Not all regular polygons can tessellate.
  b) The pentagon does not tessellate.
  c) You cannot divide 360° evenly by 108°, so the pentagon will not tessellate.
4. 60°

7.9 Tessellating Designs

When you have made your tiles, tessellate them by rotating, reflecting, and translating them. There should be no overlap or spaces between tiles in a tessellation.

Test Yourself

1. a) $A(3, 5)$, $B(-4, 1)$, $C(-3, 0)$, $D(-2, -5)$, $E(-1, 3)$, $F(0, 2)$

b)

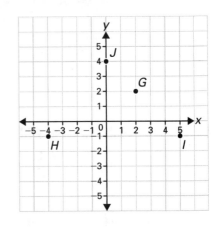

2. a) below   **b)** above
3. a) right   **b)** right
4. $A'(-5, 6)$, $B'(0, 6)$, $C'(0, 0)$, $D'(-5, 0)$
5. a)

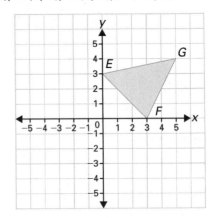

b) $E'(-5, 0)$, $F'(-2, -3)$, $G'(0, 1)$
6. $H'(-3, -2)$, $I'(1, -2)$, $J'(4, 3)$, $K'(-5, 2)$
7. a) The coordinates of the image are $M'(2, 1)$, $N'(-1, 4)$, $L'(4, 4)$.
  b) 180° ccw
8. a) A and C, B and G   **b)** E and F
9. There are several different tessellations you could draw. Make sure there is no overlap or spaces between pentominoes.
10. (1) Draw a small equilateral triangle, each side about 1 cm long. The triangle is pointing up. (2) Draw three triangles around the first triangle. Each new triangle should be congruent with the first one, and should share one side with the first triangle. Shade the three new triangles grey. (3) Together, the four triangles you have drawn make up the shape of a larger triangle pointing down. Draw a line around this larger triangle, about 2 mm away from it so it has a white border. (4) In the centre of one side of the border, draw an equilateral triangle. It should have a base about half the length of the side of the

border, and be centred in the middle of the side. Shade it grey. (5) Draw two more grey triangles, one on each side of the white border.

Chapter 8

8.1 Exploring Pattern Representations

1. a) The missing values are 3, 6, 9, 12, and 15.
   b) Start from 3. Add 3 to each value to get the next term value. An alterative rule could be: multiply the term number by 3.
2. a) The missing values are 1, 3, 4, 6, and 7.
   b) Start from 1. Add 2, then add 1, then add 2, then add 1, and so on. An alternative rule could be: Add 0, 1, 1, 2, 2, and so on, to the term numbers to get the term values.
3. b) For the 10th term, there are 30 squares in total, and 15 shaded squares.

8.2 Using Variables to Write Pattern Rules

1. a) The number of shaded squares stays the same. The number of white squares changes.
   b) Start with two shaded squares and one white square. Add one white square each time. An alternative rule could be: the total number is equal to 2 plus the term number.
   c) $2 + b$, where b is the term number
2. a) The missing values are 2, 4, 6, 8, and 10.
   b) The number of circles is equal to the term number multiplied by 2.
   c) $2c$, where c is the figure number (also called the term number)
3. a) Omar sees that the number of squares stays the same: 2. He also sees that the number of triangles is equal to the term/figure number (n) plus 1, or $n + 1$. To find the total number of blocks, Omar adds the number of squares to the number of triangles and gets $2 + (n + 1)$.
   b) Tynessa notices that the total number of blocks is equal to the term/figure number (n) plus 3. She gets $3 + n$.
4. a) $4 + s$ b) $4t$ c) $3c + 1$
5. a) n b) $3n$ c) $n + 3n$, or $4n$

6. a)

Figure number	1	2	3	4	5	...	10
Number of white squares	1	2	3	4	5		10
Number of shaded squares	8	10	12	14	16		26

b) $2n + 6$
c) $3s + 6$

8.3 Creating and Evaluating Expressions

1. a) 9, 10, 11, 12   b) 8, 16, 24, 32
   c) 7, 6, 5, 4   d) 12, 6, 4, 3
   e) 5, 7, 9, 11
2. a) 10   b) 18   c) 4   d) 6
   e) 7   f) 12   g) 0   h) 9
3. a) $17.50   b) $1.75
4. a) $21   b) $102
5. a) $65   b) $35   c) $20
6. a) $2c$ b) $10p + 2$ c) $35j - 10$
7. a) $20s + 5$ b) $65   c) $205
8. $3(x + 4) = 3(5 + 4) = 3(9) = 27$
9. a) $2p + 1$ b) 3 km   c) 9 km

8.4 Solving Equations by Inspection

1. a) 5   b) 3   c) 2
   d) 11   e) 2   f) 5
2. a) $4t - 8 = 4(2) - 8 = 0$; Ravi's solution is incorrect.
   b) $t = 6$
3. a) $2t + 1$
   b) $2t + 1 = 15$
   c) $t = 7$
   d) The figure number is $t = 10$.

8.5 Solving Equations by Systematic Trial

1. a)

Predict y.	Evaluate y + 5.	Is this the correct solution?
5	5 + 5 = 10	too low
10	10 + 5 = 15	too high
7	7 + 5 = 12	correct

b)

Predict m.	Evaluate $3m$	Is this the correct solution?
200	$3(200) = 600$	too high
150	$3(150) = 450$	too high
111	$3(111) = 333$	correct

c)

Predict r.	Evaluate $5r - 10$.	Is this the correct solution?
15	$5(15) - 10 = 65$	too low
21	$5(21) - 10 = 95$	correct
25	$5(25) - 10 = 115$	too high

2. a) $x = 64$ **b)** $q = 116$ **c)** $w = 17$ **d)** $c = 12$
  e) $e = 7$ **f)** $k = 27$ **g)** $s = 51$ **h)** $u = 31$
3. a) $4x + 100 = 140$, $x = 10$
  b) $7x = 294$, $x = 42$
  c) $4x - 52 = 212$, $x = 66$
4. a) She multiplied $(24 + 12)$ by the variable. The equation asked for multiplying the variable by 12 only, and then adding 24.
  b)

Predict z.	Evaluate $24 + 12z$.	Is this the correct solution?
10	$24 + 12(10) = 144$	too low
15	$24 + 12(15) = 204$	too high
13	$24 + 12(13) = 180$	correct

5. a) $A = 6$ units squared
  b) $h = 4$ units
  c) $b = 8$ units

8.6 Communicating the Solution for an Equation

1. On the left side there are three containers, so you get $3c$. On the right side there are 15 marbles. The equation is $3c = 15$. Divide both sides of the equation by 3, to determine that $c = 5$. The answer means that each container holds five marbles.
2. a) $5c = 10$, $c = 2$ **b)** $c + 3 = 7$, $c = 4$
  c) $4c + 5 = 13$, $c = 2$
3. On the left side, there are two containers and three marbles. You can write this as $2m + 3$. On the right side there are five marbles. The equation is $2m + 3 = 5$. Subtract 3 from both sides to get $2m = 2$. Divide both sides by 2 to get $m = 1$.
4. Tynessa should have subtracted 6 from both sides before dividing both sides by 2. The correct solution is $c = 3$.

Test Yourself

1. a) Start with one square and one triangle. Add one triangle each time. An alternative rule is: Each figure has one square and the same number of triangles as the term number.
  b) $t + 1$
2. a) $3b$ **b)** $3 + b$ **c)** $(1 + t) + 1$, or $2 + t$
3. a)

  b)

  c)

  d)

4. a) 9   **b)** 14   **c)** 2   **d)** 10
5. a) $15 + h$ **b)** $40   **c)** $115
6. a) $x = 12$ **b)** $p = 9$ **c)** $m = 2$ **d)** $b = 6$
7. a) $4 + t$ **b)** $4 + t = 16$
  c) $t = 12$ **d)** $4 + (12) = 16$
8.

Predict k.	Evaluate $4 + 2k$.	Is this the correct solution?
50	$4 + 2(50) = 104$	too low
52	$4 + 2(52) = 108$	too high
51	$4 + 2(51) = 106$	correct

9. a) $3c = 9$
  b) $c = 3$
  c) There are three containers on the left side and nine marbles on the right, so the equation is $3c = 9$. Divide both sides by 3 to get $c = 3$.
10. a) $x = 5$ **b)** $x = 3$ **c)** $x = 4$ **d)** $x = 4$

Chapter 9

9.1 Adding Fractions with Pattern Blocks

1. To show $\frac{1}{4}$ of each diagram, shade one section of the square, one section of the circle, and two sections of the rectangle.

2. For example, you could draw a rectangle divided in five equal pieces, and shade two.

3. a) To show $\frac{1}{6}$, shade one section.

b) Repeat part (a).

c) $\frac{2}{6} = \frac{1}{3}$

4. a) To show $\frac{1}{2}$, shade four sections.

b) To show $\frac{1}{8}$, shade one section. Now five sections in total are shaded.

c) $\frac{5}{8}$

9.2 Adding Fractions with Models

1. Chang forgot to convert the fraction $\frac{2}{3}$ into the equivalent fraction $\frac{4}{6}$. He should have coloured four rectangles on the first strip and one rectangle on the second strip to get a total of $\frac{5}{6}$.

2. a) $\frac{2}{4}$, or $\frac{1}{2}$ **b)** $\frac{3}{3}$, or 1

c) $\frac{6}{6}$, or 1   **d)** $\frac{13}{15}$

e) $\frac{7}{8}$ **f)** $\frac{7}{10}$

3. Draw a second arrow that is 5 units long to show $\frac{5}{20}$. Add the two arrows to get $\frac{9}{20}$.

4. a) $\frac{7}{10}$ **b)** $\frac{7}{8}$ **c)** $\frac{8}{8}$, or 1

d) $\frac{11}{24}$ **e)** $\frac{26}{40}$, or $\frac{13}{20}$ **f)** $\frac{17}{30}$

5. $\frac{3}{8}$

6. $\frac{3}{4}$ h

9.3 Multiplying a Whole Number by a Fraction

1. a) 8 squares   **b)** $\frac{8}{3}$ **c)** $2\frac{2}{3}$

2. a) $\frac{3}{4}$ **b)** $\frac{6}{5}$, or $1\frac{1}{5}$ **c)** $\frac{5}{2}$, or $2\frac{1}{2}$

d) $\frac{7}{3}$, or $2\frac{1}{3}$ **e)** $\frac{20}{6}$, or $3\frac{1}{3}$

f) $\frac{20}{7}$, or $2\frac{6}{7}$

3. a) 8   **b)** 2, 3   **c)** 3, 2   **d)** 3, 4, 2

9.4 Subtracting Fractions with Models

1. Draw a second arrow to represent $\frac{1}{4}$, or $\frac{3}{12}$. The end of the arrow should start at the tip of the first arrow, and it should point left. The tip of the second arrow will end at the solution: $\frac{5}{12}$.

2. a) $\frac{1}{4}$ **b)** $\frac{1}{10}$ **c)** $\frac{5}{21}$ **d)** $\frac{7}{30}$

e) $\frac{17}{15}$, or $1\frac{2}{15}$ **f)** $\frac{53}{28}$, or $1\frac{25}{28}$

3. $\frac{4}{15}$

9.5 Subtracting Fractions with Grids

1. a) $\frac{7}{20}$

b) Seven squares out of twenty have counters on them.

2. Jody forgot to rearrange the counters to express thirds before removing the counters in one column. Rearranging the counters leaves one full column with 2 counters left over. Removing one column leaves 2 squares out of 21 with counters. The solution is $\frac{2}{21}$.

3. a) $\frac{11}{12}$ **b)** $\frac{34}{15}$, or $2\frac{4}{15}$ **c)** $\frac{3}{8}$

d) $\frac{5}{9}$ **e)** $\frac{4}{12}$, or $\frac{1}{3}$ **f)** $\frac{5}{8}$

g) $\frac{2}{35}$ **h)** $\frac{59}{40}$, or $1\frac{19}{40}$

4. a) $\frac{17}{35}$ of the brownies

b) $\frac{18}{35}$ of the brownies

5. $\frac{17}{24}$ of the CDs

9.6 Adding and Subtracting Mixed Numbers

1. a) 4   **b)** $8\frac{1}{4}$ **c)** $7\frac{5}{6}$ **d)** $8\frac{7}{8}$

e) $16\frac{1}{12}$ **f)** $5\frac{19}{30}$ **g)** $13\frac{5}{18}$ **h)** $8\frac{13}{21}$

2. a) $2\frac{3}{4}$ **b)** $1\frac{4}{5}$ **c)** $3\frac{4}{7}$ **d)** $\frac{3}{8}$

e) $1\frac{1}{2}$ **f)** $1\frac{5}{9}$ **g)** $6\frac{1}{7}$ **h)** $\frac{1}{12}$

3. $1\frac{1}{6} + 3\frac{1}{10} = 4\frac{4}{15}$

4. a) $5\frac{1}{4}$ h   **b)** $4\frac{1}{30}$ h   **c)** $8\frac{1}{8}$ h

5. $2\frac{5}{6}$ h, or 2 h 50 min

6. a) $7\frac{1}{2}$ years old   **b)** $23\frac{1}{8}$ years old

c) $1\frac{2}{5}$ years old

7. $1\frac{1}{12}$ pizzas

8. $3\frac{5}{9}$ h

9. $1\frac{3}{40}$

9.7 Communicating about Estimation Strategies

1. a) Ryan forgot to include the fraction $\frac{9}{12}$ in his estimation. $4\frac{9}{12}$ is closer to 5 than to 4.

b) You can round off $4\frac{9}{12}$ to the number 5. Then subtract from 6. Ryan has a little more than one case of juice left over.

2. $2\frac{3}{4}$ is a little bit less than 3. $1\frac{1}{8}$ is a little bit more than 1. Add 3 and 1 to get about 4 c. of sugar in total.

3. Round off $\frac{1}{3}$ to $\frac{1}{2}$, which is easier to deal with. So the north wall needs a little less than $2\frac{1}{2}$ pieces. The west wall needs $\frac{1}{2}$ a piece. For the south wall, round off $1\frac{4}{5}$ to get a little less than 2. The east wall needs 3 pieces. Add $2\frac{1}{2} + \frac{1}{2} + 2 + 3 = 8$. Miguel needs a little less than 8 pieces of panelling.

9.8 Adding and Subtracting Using Equivalent Fractions

1. a) The common denominator is 8. The equivalent fractions are $\frac{5}{8}$ and $\frac{6}{8}$.

b) The common denominator is 10. The equivalent fractions are $\frac{5}{10}$ and $\frac{4}{10}$.

c) The common denominator is 12. The equivalent fractions are $\frac{11}{12}$ and $\frac{3}{12}$.

d) The common denominator is 35. The equivalent fractions are $\frac{20}{35}$ and $\frac{28}{35}$.

2. a) The missing values are 2 and 1.
b) The missing values are 6, 10, 16, and 1.

3. a) $\frac{11}{14}$ **b)** $\frac{5}{8}$ **c)** $\frac{14}{9}$, or $1\frac{5}{9}$

d) $\frac{31}{42}$ **e)** $\frac{13}{8}$, or $1\frac{5}{8}$ **f)** $\frac{37}{40}$

g) $\frac{9}{20}$ **h)** $\frac{5}{24}$

4. a) $\frac{1}{6}$ **b)** $\frac{7}{20}$ **c)** $\frac{5}{10}$ or $\frac{1}{2}$

d) $\frac{1}{72}$ **e)** $\frac{5}{14}$ **f)** $\frac{5}{12}$ **g)** $\frac{2}{35}$

h) $\frac{27}{60}$, or $\frac{9}{20}$

5. $1\frac{7}{12}$ h

6. $2\frac{1}{3}$ days

7. Indira drank $\frac{1}{15}$ of a bottle more lemonade than Simon.

8. $1\frac{1}{7}$ h

9. Jody has finished $\frac{8}{35}$ more of her homework than Sandra.

10. a) $\frac{3}{10}$ **b)** Colin won $\frac{2}{5}$ more than Kaitlyn.

Test Yourself

1. a) C **b)** A **c)** B

2. a) $\frac{3}{4}$ **b)** $\frac{3}{4}$ **c)** $\frac{2}{5}$

d) $1\frac{1}{2}$, or $\frac{3}{2}$ **e)** $3\frac{2}{5}$, or $\frac{17}{5}$

f) $4\frac{1}{5}$, or $\frac{21}{5}$

3. a) $\frac{7}{6}$, or $1\frac{1}{6}$ **b)** $\frac{1}{10}$

4. a) $\frac{1}{9}$ **b)** $\frac{17}{24}$

5. a) 1 **b)** $\frac{1}{2}$ **c)** $\frac{1}{10}$ **d)** $\frac{7}{24}$

e) $\frac{7}{9}$ **f)** $\frac{3}{14}$ **g)** $\frac{43}{40}$, or $1\frac{3}{40}$

h) $\frac{11}{15}$

6. a) $3\frac{1}{3}$ **b)** $2\frac{1}{10}$

7. $\frac{1}{6}$

8. a) $4\frac{5}{6}$ **b)** $11\frac{11}{12}$ **c)** $9\frac{31}{56}$

9. a) $1\frac{3}{5}$ **b)** $1\frac{5}{6}$ **c)** $4\frac{1}{10}$

10. a) $\frac{3}{8}$ **b)** $\frac{1}{9}$ **c)** $\frac{1}{18}$

11. $\frac{2}{3}$ of her pay

12. a) $75\frac{7}{10}$ years old **b)** $79\frac{11}{12}$ years old

c) $72\frac{3}{10}$ years old

13. a) 1 full box **b)** $\frac{1}{4}$ of a box

14. $\frac{7}{12}$

15. a) $\frac{7}{8}$ of a tube is a little less than 1 tube. $3\frac{1}{6}$ tubes is a little more than 3 tubes. Added together, Kaitlyn used about 4 tubes of paint.

b) She has about 5 tubes of paint left in total.

16. $\frac{7}{10}$ of the day

Chapter 10

10.1 Building and Packing Prisms

1. a) 3 sides, 3 vertices
   b) 9 edges, 5 faces, 6 vertices
2. a) 12 edges, 6 faces, 8 vertices
   b) 15 edges, 7 faces, 10 vertices

10.2 Building Objects from Nets

1. (a), (d), and (e)
2. a) C   b) A and C   c) B   d) B
3. a) It will look like a cube with a square pyramid on top.

   b)

4. a)

   b)

   c)

10.3 Top, Front, and Side Views of Cube Structures

1. a)
   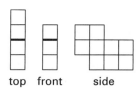

   top   front   side

   b)
   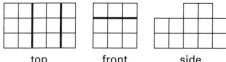

   top   front   side

   c)
   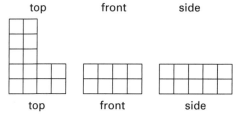

   top   front   side

d)
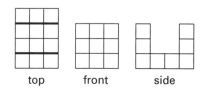

top   front   side

10.4 Top, Front, and Side Views of 3-D Objects

1. rectangular prism
2. hexagonal prism
3.

   top   front   side

4.

   top   front   side

5.

   top   front   side

10.5 Isometric Drawings of Cube Structures

1. a)   b)   c)

10.6 Isometric Drawings of 3-D Objects

1. a)

   b)

10.7 Communicating about Views

1. a) Follow these steps to draw top, front, and side views of a pyramid with a square base. (1) Draw the pyramid. (2) Label the front and sides. (3) Imagine you are looking at the pyramid from above. You would see the square base, the tip of the pyramid, and the edges of the pyramid

going from the tip to each vertex of the base. Draw what you see using thick black lines to show changes in depth.
(4) Imagine you are looking at the pyramid from the front. You would see only the face of a triangle. Draw what you see.
(5) Imagine you are looking at the pyramid from one side. You would see only the face of a triangle. Draw what you see.

b)

top	front	side

2. Follow these steps to build a structure using the top, front, and side views. (1) Start with the side view, because there are no changes in depth. Use six cubes to build the side. (2) Look at the top view of the structure you have built. It matches the top view in the drawing. (3) Look at the front view of the structure you have built. It matches the front view in the drawing. (4) Your structure is finished, because it matches all three views.

Test Yourself

1. C
2. A and B
3.

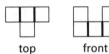

top	front	side

4.

5. **a)**

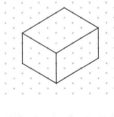

b)

6. Follow these steps to draw top, front, and side views of the structure. (1) Imagine you are looking at the structure from above. You will see six cubes arranged in a rectangle with a width of two cubes and a length of three cubes. The top left cube is raised above the rest. (2) Imagine you are looking at the structure from the front. There are two cubes side by side. Directly above the cube on the left is another cube at a different depth. (3) Imagine you are looking at the structure from the right. There are three cubes in a row. Directly above the cube on the far right is another cube at a different depth.

Chapter 11

11.1 Surface Area of a Rectangular Prism

1. **a)** 4 cm by 6 cm; 24 cm²
   b) 4 cm by 2 cm; 8 cm²
   c) 2 cm by 6 cm; 12 cm²
   d) Surface Area = 2 × (24 cm² + 8 cm² + 12 cm²) = 88 cm²
2. **a)** 28 cm² **b)** 32 cm² **c)** 78 cm²
3. **a)** 46 cm² **b)** 280 cm² **c)** 138.2 cm²

11.2 Volume of a Rectangular Prism

1. **a)** 12 cubes   **b)** 32 cubes
   c) 120 cubes
2. **a)** 64 cm³   **b)** 1176 cm³
3. **a)** 12 cm³   **b)** 450 cm³
   c) 64 cm³   **d)** 196 cm³
   e) 100 000 cm³   **f)** 1237.5 cm³
4. She should buy (c), the Super Clean filter.
5. **a)** 4 cm   **b)** 4 cm   **c)** 9 cm   **d)** 260 cm³
   e) 8 cm   **f)** 12 cm
6. Two possible sets of dimensions are 10 m by 3 m by 2 m; and 6 m by 5 m by 2 m.
7. Brand 3 is the best buy, because it has the lowest price per cm³. (Find the volume, then divide the price by the volume to find the price per cm³.)
8. 128 units³
9. Yes, Miguel should accept. Jody's box has a larger volume than his box.

11.3 Solve Problems by Guessing and Testing

1. **b)** 2 × 6 × 10   **c)** 12 × 5 × 2
   d) 6 × 5 × 4   **e)** 4 × 15 × 2

2. b)

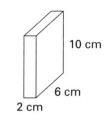

c)

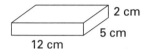

d)

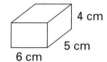

e)

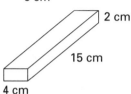

3. The three prisms can have dimensions 2 cm × 9 cm × 10 cm; 3 cm × 6 cm × 10 cm; 6 cm × 6 cm × 5 cm; or any other combination of three numbers that multiply to give 180 cm³.

4.

Length (cm)	Width (cm)	Height (cm)	Surface Area (cm²)
100.0	40.0	50.0	22 000
90.0	44.4	50.0	21 432
80.0	50.0	50.0	21 000
70.0	57.1	50.0	20 704
60.0	66.7	50.0	20 674
50.0	80.0	50.0	21 000

According to the chart above, a base with length = 60.0 cm and width = 66.7 cm results in the smallest surface area. This is a good answer. However, you can keep going to find a better answer. Notice that these dimensions are almost equal. From this, you can guess that a length and width that are equal will result in the smallest possible surface area:

Length (cm)	Width (cm)	Height (cm)	Surface Area (cm²)
63.2	63.2	50	20 628.5

5. 25 cm, 30 cm, and 10 cm

11.4 Relating the Dimensions of a Rectangular Prism to Its Volume

1. **a) i)** 4 cm × 6 cm × 10 cm
   ii) 4 cm × 5 cm × 12 cm

b) If you doubled the height of the dimensions in part (i), the new volume would be 480 cm³, or double the original volume.

c) Yes, the new volume would be equal. Doubling any one dimension results in a volume that is doubled.

d) i) 480 cm³ **ii)** 480 cm³

2. **a)** 720 cm³ **b)** 180 cm³ **c)** 1080 cm³
   d) 5 cm **e)** 100 cm **f)** 1 cm

11.5 Exploring the Surface Area and Volume of Prisms

1. **a)** 96 cm² and 64 cm³
   b) 112 cm² and 64 cm³
   c) 136 cm² and 64 cm³
2. **a)** 64.0 cm² and 28.0 cm³
   b) 64.0 cm² and 32.0 cm³
   c) 64.0 cm² and 34.848 cm³
3. The prism on the right side has the greatest surface area. If two prisms have the same volume, the prism that is closest in shape to a cube will have the smallest surface area.
4. The prism on the left side has the greatest volume. If two prisms have the same surface area, the prism that is closest in shape to a cube will have the greatest volume.

Test Yourself

1. **a)** 78 units²   **b)** 32 units²
   c) 142 cm²
2. **a)** 24 units³   **b)** 48 units³
   c) 360 cm³
3. **a)** 248 cm²   **b)** 240 cm³
   c) 120 cm³   **d)** 480 cm³
4. **a)** 30 cm³   **b)** 1 cm
   c) 3 cm   **d)** 125 cm³
   e) 21 cm³
5. Sandra's tower should be 3 blocks high.

Chapter 12

12.1 Exploring Probability

1. **a)** probably $\frac{1}{2}$ to 1, depending on your habits
   b) $\frac{1}{2}$ **c)** 0   **d)** 0
   e) $\frac{1}{2}$

2. **a)** This is not a fair game.
   b) Omar is most likely to win.

3. a) red marble, red marble, blue, blue, blue, yellow, yellow, green, green, green, green green

b) $\frac{3}{12}$ or $\frac{1}{4}$

12.2 Calculating Probability

1. a) $\frac{1}{3}$ **b)** $\frac{1}{3}$

2. a) red + green, red + blue, green + red, green + blue, blue + red, blue + green
   b) 0.333

3. $\frac{6}{30}$, or $\frac{1}{5}$

4. a) $\frac{1}{4}$ **b)** $\frac{3}{4}$

12.3 Solve Problems Using Organized Lists

1. a)

$5 bills	3	2	2	1	1	1	0	0	0	0
$10 bills	0	1	0	2	0	1	3	0	2	1
$20 bills	0	0	1	0	2	1	0	3	1	2
Sum	$15	$20	$30	$25	$45	$35	$30	$60	$40	$50

b) 10 different combinations are possible
c) 1 combination adds up to $60

d) $\frac{1}{10}$ **e)** $\frac{7}{10}$

2. a)

Win	4	3	3	2	2	2	1	1	1	1	0	0	0	0	0
Lose	0	1	0	2	0	1	3	0	2	1	4	0	3	1	2
Tie	0	0	1	0	2	1	0	3	1	2	0	4	1	3	2

b) $\frac{1}{15}$

12.4 Using Tree Diagrams to Calculate Probability

1. a)

Bread		Jam	Outcome
white	→	blueberry	WB
	→	strawberry	WS
	→	raspberry	WR
whole wheat	→	blueberry	WWB
	→	strawberry	WWS
	→	raspberry	WWR
pumpernickel	→	blueberry	PB
	→	strawberry	PS
	→	raspberry	PR

b) $\frac{3}{9}$, or $\frac{1}{3}$ **c)** $\frac{1}{9}$

2. a)

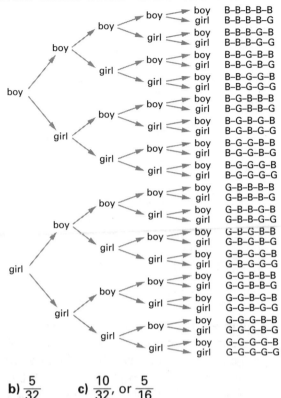

1st child	2nd child	3rd child	4th child	5th child	Outcome

B–B–B–B–B
B–B–B–B–G
B–B–B–G–B
B–B–B–G–G
B–B–G–B–B
B–B–G–B–G
B–B–G–G–B
B–B–G–G–G
B–G–B–B–B
B–G–B–B–G
B–G–B–G–B
B–G–B–G–G
B–G–G–B–B
B–G–G–B–G
B–G–G–G–B
B–G–G–G–G
G–B–B–B–B
G–B–B–B–G
G–B–B–G–B
G–B–B–G–G
G–B–G–B–B
G–B–G–B–G
G–B–G–G–B
G–B–G–G–G
G–G–B–B–B
G–G–B–B–G
G–G–B–G–B
G–G–B–G–G
G–G–G–B–B
G–G–G–B–G
G–G–G–G–B
G–G–G–G–G

b) $\frac{5}{32}$ **c)** $\frac{10}{32}$, or $\frac{5}{16}$

12.5 Applying Probabilities

1. Romona is the most likely to make her next shot.
2. a) Indira is the most likely to win.
   b) Bonnie and Simon have the same probability of winning.

3. a) $\frac{1}{4}$ **b)** $\frac{1}{6}$

   c) Paul is the most likely to win.

4. $\frac{1}{3}$

5. a) Bag B   **b)** $\frac{12}{17}$

6. James's throw of 2 was the least likely event.
7. The most likely total is 7, because you can get it in the most number of ways (1 + 6, 3 + 4, and 2 + 5).
8. It is not certain that she has touched a Norway maple, although it is very likely. Calculating the probabilities will show that there is a chance of touching three trees in a row that are not Norway maples.

Test Yourself

1. a) $\frac{2}{3}$ **b)** $\frac{1}{2}$

2. a) $\frac{3}{10}$ **b)** $\frac{2}{10}$ **c)** $\frac{1}{10}$ **d)** $\frac{2}{5}$

3. a)

Quarters	4	3	3	2	2	2	1	1	1	1	0	0	0	0	0
Dimes	0	1	0	2	0	1	3	0	2	1	4	0	3	1	2
Nickels	0	0	1	0	2	1	0	3	1	2	0	4	1	3	2
Total value	100	85	80	70	60	65	55	40	50	45	40	20	35	25	30

b) $\frac{1}{15}$

c) You are certain to guess them, so the probability is 1, or 100%.

4. a)

Story 1	1	1	1	1	1	2	2	2	2	3	3	3	4	4	5
Story 2	1	5	2	4	3	1	4	2	3	1	3	2	1	2	1
Story 3	5	1	4	2	3	4	1	3	2	3	1	2	2	1	1

b) $\frac{1}{5}$ **c)** $\frac{2}{5}$

5. a)

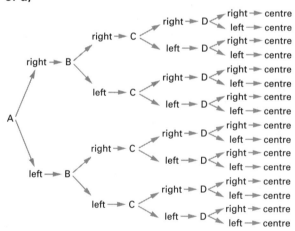

b) $\frac{1}{16}$ **c)** $\frac{3}{8}$

6.

Shoes	Jeans	Sweaters
1st pair	1st pair	1st, 2nd, 3rd
	2nd pair	1st, 2nd, 3rd
2nd pair	1st pair	1st, 2nd, 3rd
	2nd pair	1st, 2nd, 3rd

7. a) Team 2 is the most likely to win.
  b) Team 1 is the most likely to lose.

8. a) Romona should choose Tynessa's wallet.

  b) $\frac{6}{25}$